Shape Selective Catalysis in Industrial Applications

CHEMICAL INDUSTRIES

A Series of Reference Books and Textbooks

Consulting Editor
HEINZ HEINEMANN
Heinz Heinemann, Inc.
Berkeley, California

Shape Selective Catalysis in Industrial Applications

N. Y. Chen

Mobil Research and Development Corporation
Princeton, New Jersey

William E. Garwood and Francis G. Dwyer

Mobil Research and Development Corporation
Paulsboro, New Jersey

Marcel Dekker, Inc. • **New York and Basel**

o 3436 433
CHEMISTRY

Library of Congress Cataloging-in-Publication Data

Chen, N. Y.
    Shape selective catalysis in industrial applications / N. Y. Chen,
William E. Garwood, and Frank G. Dwyer.
      p. cm. -- (Chemical industries ; v. 36)
    Includes bibliographies and index.
    ISBN 0-8247-7856-1
    1. Catalysis. 2. Zeolites. I. Garwood, William E.
II. Dwyer, Francis G. III. Title. IV. Series.
TP156.C35C459 1989
660.2'995--dc19   88-32354
                                               CIP

This book is printed on acid-free paper.

MARCEL DEKKER, INC.
270 Madison Avenue, New York, New York 10016

Current printing (last digit):
10 9 8 7 6 5 4 3 2 1

PRINTED IN THE UNITED STATES OF AMERICA

Preface

In the course of my 29-year career with Mobil, I have been fortunate enough to witness and play a role in the evolution of shape selective catalysis from a relatively straightforward concept into a major branch of industrial catalytic process technology. During this period, many new zeolite-based processes have entered into commercial practice in the petroleum, petrochemical, and chemical industries. The majority of these have been based on the shape selective attributes of these zeolites. With the technology of catalytic shape selectivity having reached a certain degree of maturity, it seemed an appropriate juncture to write about the principles of shape selective catalysis and the industrial application of shape selective zeolites.

The idea for this undertaking was started over ten years ago at the urging of Heinz Heinemann, the founding editor of *Catalysis Reviews* and an old colleague, now at the University of California—Berkeley. Bill Garwood and I, together with Werner Haag and Al Schwartz, wrote two general review papers on the subject of hydrocarbon conversion over shape selective zeolite catalysts. These papers were presented at the First Symposium on Advances in Catalytic Chemistry in Snowbird, Utah, and at the AIChE 72nd Annual Meeting in San Francisco, California, in 1979.

It took Bill and me more than five years to complete the next phase of our project, which was to review the rapidly expanding number of journal articles and patents describing the industrial applications of shape selective catalysis. This work was published in *Catalysis Reviews—Science and Engineering* in 1986. The present work is an updated and greatly expanded version of this review paper. We've had the good fortune of adding Frank Dwyer as a coauthor. Frank, who has over 30 years of experience in the area of zeolite catalysis, has contributed significantly to the content of the first four chapters.

This book is divided into nine chapters. Introductory material, including a brief presentation of the fundamentals of zeolite catalysis, constitutes the first three chapters. We hope that these chapters will be useful to the reader in bridging the gap between the basic principles of shape selective catalysis and their practical applications.

The major portion of this book, Chapters 4–8, is devoted to shape selective reactions catalyzed by zeolites and their industrial applications. Attempts have been made to include in Chapter 4 all the hydrocarbon and non-hydrocarbon reactions catalyzed by shape selective zeolites. Chapters 5 and 6 describe commercialized or developed processes in the petroleum and petrochemical industries, respectively. Chapter 7 describes the new processes for the production of fuels and chemicals from nonconventional feedstocks. Included in this chapter is an updated description of Mobil's MTG process for the production of gasoline from methanol, a process that has already been commercialized in New Zealand. Chapter 8 covers a diverse field of potential opportunities for the application of shape selective catalysis including: oil production, shale oil, coal, natural gas upgrading, internal combustion engine modification, biomass conversion, and applications in the fermentation, chemical, and waste recovery industries.

Judging from the steady increase in the number of zeolite process application patents filed each year in the U.S. Patent Office and patent offices around the world, there is little doubt that the application of shape selective catalysis will continue to grow and extend to metal and bifunctional catalyzed reactions

as well as to the commercialization of non-hydrocarbons processes
and the synthesis of fine chemicals. Frequently, discoveries from
one industry fertilize new developments in other, seemingly un-
related industries. Thus, it is the fervent hope of the authors
that this book will serve as a bridge of "technology transfer"
among various industries. Most importantly, however, we hope
that this book will serve as a useful reference for workers in the
field.

The advent of shape selective catalysis can be traced back
to the discovery of the unique catalytic selectivity of Zeolite A
by Paul Weisz and Vince Frilette at Mobil in the late 1950s.
The discovery, although exciting to a few of us, did not set the
world on fire. In fact, at the time there was not a zeolite that
could withstand the rigors of the industrial environment and be
considered of practical value. A small group of zeolite enthusi-
asts labored for five years in relative obscurity before making a
practical breakthrough, which led to the commercialization of
the first catalytic process, the Selectoforming process, using a
stable shape selective zeolite as the catalyst.

We at Mobil owe a debt of gratitude to my mentor, Paul
Weisz, for his conviction and his foresight in emphasizing the
significance of shape selective catalysis. He reminded us of the
similarity of shape selective catalysis to enzyme catalysis, where
the geometric shape of a molecule and its environment play important roles. Admittedly, the activity and specificity of shape
selective zeolite catalysts are still orders of magnitude less than
those of enzyme catalysts.

It is also to the credit of Mobil's management to have taken
the risk, marshalling at times the entire staff of two laboratories,
for the development of some of these processes described in this
book. The announcement of the MTG process in 1976 caught
the imagination of the academic world and stimulated fundamental research on medium-pore zeolites and C_1 chemistry.
This is evidenced by the explosive proliferation of technical papers
on these subjects, which, in turn, has greatly expanded our knowledge base. Hopefully, a careful study of where shape selective
zeolite catalysis has been applied will lead to additional industrial
applications.

On behalf of my coauthors, Bill Garwood and Frank Dwyer, I wish to thank many of our colleagues who took the time to review and comment on the manuscript. Mrs. Delma Dow, Mrs. Diane Tavin, and Mrs. Marianne Snyder deserve special recognition for undertaking the administrative details associated with preparing the manuscript.

N. Y. Chen

Contents

Shape Selective Catalysis in Industrial Applications

1
Introduction

Nearly 30 years ago, Weisz and Frilette (1960) coined the term "shape-selective catalysis" to describe the then unexpected intrinsic catalytic activities of synthetic crystalline molecular sieve zeolites. They found that the calcium ion exchanged zeolite A having "port" sizes of 4 to 5 Å selectively cracked straight chain n-paraffins to exclusively straight chain products. Since that time, great strides in the use of shape selective catalysts in commercial catalytic processes have been made. Their applications have expanded well beyond the boundaries of traditional petroleum refining to petrochemical and chemical manufacturing.

Over the past 30 years, many new synthetic zeolites have been discovered. Of particular importance was the discovery of the synthetic medium-pore zeolites with "port" sizes of 5 to 6 Å. These include, among others, ZSM-5 (Argauer and Landolt, 1972), ZSM-11 (Chu, 1973), ZSM-23 (Plank et al., 1978; Valyocsik, 1984b), ZSM-35 (Plank et al., 1977), ZSM-48 (Chu, 1983;

1

Rollmann and Valyocsik, 1983), NU-6 (Whittam, 1983), and
Theta-1 (Ball et al., 1985).*

While classical zeolites contain Si and Al, the addition of
the elements Ga, Ge, Be, B, Fe, Cr, P, and Mg in framework
positions has also been achieved. Indeed, the number of elements that can be inserted into the crystal framework either by
isomorphous substitution or by direct synthesis has grown dramatically in recent years (Flanigan et al., 1986).

The availability of synthetic zeolites has greatly expanded
the realm of "shape selectivity." Beginning with the original
discovery of selective conversion of straight chain molecules, it
now has become possible to selectively convert such molecules
as certain branched molecules, single-ring aromatics, naphthenes,
and nonhydrocarbons with a critical molecular dimension less
than about 6 Å.

During this period, major technological advances have also
resulted from research and development efforts at Mobil and
other research laboratories. The commercial production of
ZSM-5, the commercialization of new petroleum refining and
petrochemical processes, and the ZSM-5 catalyzed methanol-to-
gasoline (MTG) process have attracted worldwide attention.
Interest in medium-pore zeolites has grown in recent years, both
at academic institutions and at other industrial laboratories. This
is evidenced by the exponential growth of publications dealing
with the fundamental and applied aspects of shape selective
catalysis. Molecular shape selective zeolites have emerged from
being laboratory curiosities to becoming commercially significant
groups of industrial catalysts.

Over the years, the principles of shape selective catalysis
have been comprehensively reviewed by a number of authors

---

*Not listed are the many zeolite compositions that have been given different
nomenclature but are topologically isostructural and otherwise the same as
some of the above zeolites, for example, Silicalite, AMS-1, AZ-1, CZM,
FZ-1, NU-5, TSZ's, TZ's, ZBM's, ZMQ's and ZSM-5; ISI-1, KZ-2, NU-10,
ZSM-22 and Theta-1.

(Venuto, 1968; Weisz, 1973; Csicsery, 1976; Chen et al., 1979a,b; Weisz, 1980; Derouane, 1980; Heinemann, 1981; Derouane, 1982; Whyte and Dalla Betta, 1982; Dwyer and Dyer, 1984; Haag, 1984a,b; Csicsery, 1984; and Csicsery, 1986). Our purpose here is not to review the subject matter in general, but to summarize the impact of shape selective catalysis on the petroleum and petrochemical industries. Special emphasis will be placed on medium-pore zeolites.

To provide a brief introduction to the fundamentals of zeolite catalysis, we have included two short chapters that describe the relation between catalyst structure and catalytic activity, and the principal methods of achieving molecular shape selectivity. We hope these chapters will be useful to the reader in bridging the gap between the basic principles of shape selective catalysis and their practical applications. The principal emphasis of this review can be found in Chapters 4 through 8.

REFERENCES

Argauer, R. J. and G. R. Landolt, U.S. Pat. 3,702,866, Nov. 14, 1972.

Ball, W. J., S. A. I. Barri, and D. Young, U.S. Pat. 4,533,649, Aug. 6, 1985 (U.K. Appl. 8225524, Sep. 7, 1982).

Chen, N. Y., W. E. Garwood, W. O. Haag, and A. B. Schwartz, "Shape Selective Hydrocarbon Catalysis Over Synthetic Zeolite ZSM-5," paper presented at Symp. Advan. Catal. Chem. I, Snowbird, Utah, Oct. 3, 1979a.

Chen, N. Y., W. E. Garwood, W. O. Haag, and A. B. Schwartz, "Shape Selective Conversion Over Intermediate Pore Size Zeolite Catalysts," paper presented at the AIChE 72nd Annual Mtg., San Francisco, Nov. 25-29, 1979b.

Chu, P., U.S. Pat. 3,709,979, Jan. 9, 1973.

Chu, P., U.S. Pat. 4,397,827, Aug. 9, 1983.

Csicsery, S. M., *Zeolite Chemistry and Catalysis*, J. A. Rabo, ed., Am. Chem. Soc. Monogr. *171*, 680 (1976).

Csicsery, S. M., Zeolites *4*, 202 (1984).

Csicsery, S. M., Pure Appl. Chem. *58*, 841 (1986).

Derouane, E. G., Stud. Surf. Sci. Catal. *5*, 5 (1980).

Derouane, E. G., "Intercalation Chemistry," W. M. Stanley and A. J. Jacobson, eds., Academic Press, New York, p. 101 (1982).

Dwyer, J. and A. Dyer, Chem. Ind. *265*, 237 (1984).

Flanigan, E. M., B. M. Lok, R. L. Patton, and S. T. Wilson, Proc. 7th Int. Zeol. Conf., Y. Murakami, A. Iijima, and J. W. Ward, eds., Kodansha/Elsevier, Tokyo/Amsterdam, p. 103 (1986).

Haag, W. O., *Heterogeneous Catalysis*, Vol. 2, B. L. Shapiro, ed., Texas A & M University Press, College Station, Tex., p. 95 (1984a).

Haag, W. O., Proc. 6th Int. Zeol. Conf., Reno, Nevada, D. Olson and A. Bisio, eds., Butterworths, Surrey, U.K., p. 466 (1984b).

Heinemann, H., Catal. Rev.-Sci. Eng. *23*, 315 (1981).

Plank, C. J., E. J. Rosinski, and M. K. Rubin, U.S. Pat. 4,016,245, Apr. 5, 1977.

Plank, C. J., E. J. Rosinski, and M. K. Rubin, U.S. Pat. 4,076,842, Feb. 28, 1978.

Rollmann, L. D. and E. W. Valyocsik, U.S. Pat. 4,423,021, Dec. 27, 1983.

Valyocsik, E. W., U.S. Pat. 4,481,177, Nov. 6, 1984a.

Valyocsik, E. W., U.S. Pat. 4,490,342, Dec. 25, 1984b.

Venuto, P. B. and P. S. Landis, Advan. Catal. *18*, 259 (1968).

Weisz, P. B. and V. J. Frilette, J. Phys. Chem. *64*, 382 (1960).

Weisz, P. B., Chemtech *3*, 498 (1973).

Weisz, P. B., Pure Appl. Chem. *52*, 2091 (1980).

Whittam, T. V., U.S. Pat. 4,397,825, Aug. 9, 1983 (U.K. Appl. 8039685, Dec. 11, 1980).

Whyte, T. E. and R. A. Dalla Betta, Catal. Rev.-Sci. Eng. *24*, 567 (1982).

2
Relation Between Zeolite Structure and Its Catalytic Activity

I. ZEOLITES

For decades, zeolites have been described as crystalline alumino-silicate molecular sieves, which have open porous structures and ion exchange capacities. More recently, we have become aware that these materials may contain elements in addition to silicon and aluminum in their framework structures.

Nature has provided us with 34 different zeolites (Barrer, 1968; Breck, 1974; Meier and Olson, 1978; Meier, 1979). But among those of interest to catalysis, only a few are found in abundance and even fewer have found industrial use. The industrial application of zeolite catalysts depends largely on our ability to synthesize zeolites, and the synthesis of known and new structures has made new discoveries in zeolite catalysis possible. Today, more than 50 different aluminosilicate zeolite structures are available (von Ballmoos, 1984), with pore openings that allow the passage of molecules ranging in size from less than 5 Å to greater than 10 Å.

Many crystalline metallophosphate zeolites, either having the same framework structures as aluminosilicates or new structures, have been recently synthesized (Wilson et al., 1982a; 1982b; Lok et al., 1984; Lok et al., 1985; Wilson and Flanigan, 1986).

The catalytic sites in aluminosilicate zeolites are associated with tetrahedral aluminum atoms in substitutional positions in the framework of silica. In the case of hydrogen zeolites, protons associated with the negatively charged framework aluminum are the source of Brønsted acid activity.

Metallophosphate zeolites, on the other hand, derive their acid activity by the isomorphous substitution in the

aluminophosphate framework structure. With equimolar concentration of the trivalent aluminum and pentavalent phosphorus,
the parent structure, like the tetravalent silicon in silica, has no
intrinsic acidity. An imbalance of aluminum and phosphorus
through the replacement of either aluminum or phosphorus by
other atoms, such as silicon, results in a charged framework and
a source of acidity or basicity.

II. PORE/CHANNEL SYSTEMS

Zeolites of interest to shape selective catalysis may be divided
into three major groups according to their pore/channel systems.
Some of their structural properties are presented in Table 2.1.

A. 8-Membered Oxygen Ring Systems

These include most of the earliest known shape selective small
pore zeolites, such as Linde A, erionite, and chabazite. Other
members of this group include ZK-5, ZSM-34, and high silica
analogs of Linde A, that is, zeolite alpha, ZK-4, ZK-21, and
ZK-22, and several other less common natural zeolites.

 Among the many metallophosphate zeolites (ALPO$_4$, SAPO,
TAPO, MeAPO, etc.) 14, 17 (erionite), 18, 26, 33, 34 (chabazite), 35 (levynite), 39, 42 (zeolite A), 43 (gismondine), 44, and
47 are believed to be small pore materials.

 The shape of the 8-membered oxygen rings varies from
circular to puckered and elliptical. The dimension of the pore
opening also varies accordingly. For example, Linde A and
ZK-4 have circular openings, while erionite and chabazite have
puckered and elliptical pore openings.

 Members of this group sorb straight chain molecules, such
as n-paraffins and olefins and primary alcohols. It is noted that
these molecules have critical dimensions larger than the pore
size values derived from crystallographic data. Table 2.2 shows
such a comparison. The reason for this apparent incongruity is

Table 2.1   Pore Structure of Zeolites

Type code	Name	Pore system	Pore dimensions (Å)
BIK	Bikitaite	8	3.2 × 4.9
BRE	Brewsterite	8	2.3 × 5.0
CHA	Chabasite	8	3.6 × 3.7
DAC	Dachiardite	10; 8	3.7 × 6.7; 3.6 × 4.8
EAB	TMA-E(AB)	8	3.7 × 4.8
EDI	Edingtonite	8	3.5 × 3.9
EPI	Epistilbite	10; 8	3.2 × 5.3; 3.7 × 4.4
ERI	Erionite	8	3.6 × 5.2
FAU	Faujasite(X,Y)	12	7.4
FER	Ferrierite	10; 8	4.3 × 5.5; 3.4 × 4.8
GME	Gmelinite	12; 8	7.0; 3.6 × 3.9
HEU	Heulandite	10; 8	4.0 × 5.5; 4.4 × 7.2
KFI	ZK-5	8	3.9
LTA	Linde Type A	8	4.1
LTL	Linde Type L	12	7.1
MAZ	Mazzite	12	7.4
MEL	ZSM-11	10	5.1 × 5.5
MFI	ZSM-5	10	5.4 × 5.6; 5.1 × 5.5
MOR	Mordenite	12; 8	6.7 × 7.0; 2.9 × 5.7
OFF	Offretite	12; 8	6.4; 3.6 × 5.2
PAU	Paulingite	8	3.9
RHO	Rho	8	3.9 × 5.1
STI	Stilbite	10; 8	4.1 × 6.2; 2.7 × 5.7

Source: Meier and Olson (1978).

Table 2.2   Comparison of Pore
Size and Critical Dimension of
Sorbate Crystallographic Pore
Size, Å

Linde 5A	4.1
Erionite	3.6 × 5.2
Chabazite	3.6 × 3.7
Molecular dimensions, Å[a]	
n-Hexane	3.9 × 4.3 × 9.1

[a]Estimated from Courtald space
filling models.
Source: Data taken from Meier and
Olson (1978).

that these dimensions are calculated on the basis of hard spheres;
in reality, the effective pore size depends on the interaction of
the intramolecular and the interatomic forces between the host
structure and the diffusing molecules. Thus the actual pore size
can be significantly larger than that calculated (Reischman and
Olson, 1986; Wu and Landolt, 1986).

    The pore/channel systems of these zeolites also contain inter-
connecting "supercages," which are much larger than the size of
the connecting "windows." The supercage/window structure is
often blamed for the cause of catalyst deactivation or coking of
acidic catalysts. Bulky molecules such as polycyclic aromatics,
which are formed in the supercages, cannot escape through the
windows. These molecules are trapped within and end up as
coke deposits (Venuto et al., 1966; Venuto and Landis, 1968;
Rollmann, 1977; Walsh and Rollmann, 1977, Walsh and Rollmann, 1979; Rollmann and Walsh, 1979; Rollmann and Walsh,
1982).

    The first zeolite used in a commercial molecular shape selec-
tive catalytic process, the Selectoforming process, (see Chapter 5)
was one of these small pore zeolites, erionite. Figure 2.1 shows

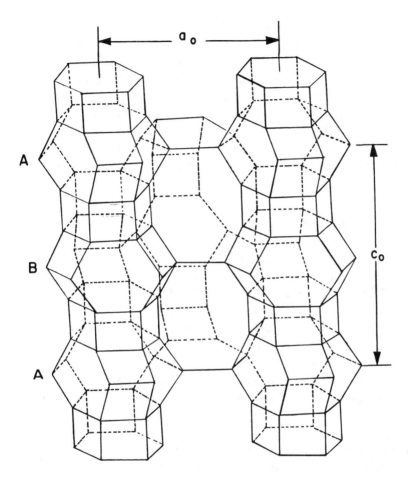

Figure 2.1   Erionite framework. a_0 and c_0, principal crystallographic axes; ABA, stacking order of cancrinite cages. (From Gorring, 1973.)

the framework structure of erionite (Staples and Gard, 1959;
Bennett and Gard, 1967). The zeolite has both large 12-mem-
bered oxygen rings and small 8-membered oxygen rings in its
framework structure. The large 12-membered oxygen ring
channels are formed by bridging columns of cancrinite cages
with hexagonal rings. However, these channels are blocked in
their longitudinal directions. The only access to these channels
are 8-membered oxygen ring openings that are perpendicular to
the channels. Therefore, erionite belongs to the 8-membered
oxygen ring system, and it sorbs only straight chain molecules.

B. 10-Membered Oxygen Ring Systems

These are also known as medium pore zeolites. Among the
varieties of unique crystal structure types in this group are
ZSM-5 (Kokotailo et al., 1978a) and ZSM-11 (Kokotailo et al.,
1978b), which are known as pentasils,* Theta-1 (Barri et al.,
1984), which is isostructural with ZSM-22 (Kokotailo et al.,
1985), ZSM-23 (Rohrman et al., 1985), ZSM-48 (Schlenker et
al., 1985), and laumontite.

Except laumontite, which has puckered 10-membered
oxygen rings in its structure, almost all medium pore zeolites of
interest to shape selective catalysis are synthetic in origin. Their
framework structures contain 5-membered oxygen rings and
they are more siliceous than previously known zeolites. In
many instances, these zeolites may be synthesized with a pre-
dominance of silicon and with only a very small concentration
of aluminum and other atoms. Thus, these zeolites may be considered as "silicates" with framework substitution by small
quantities of aluminum and other elements (Dwyer and Jenkins,
1976).

---

*The term pentasil was defined by Kokotailo and Meier (1980) as a family
of zeolites having similar structures with ZSM-5 and ZSM-11 as its two end
members. These framework structures are all formed by linking chains of
5-membered ring secondary building units.

Among the metallophosphate zeolites ($ALPO_4$, SAPO, TAPO, MeAPO, etc.) 11, 31, 41 are believed to have unidimensional medium pore structures.

As in the case of the 8-membered oxygen ring zeolites, the shape and size of the 10-membered oxygen rings also varies from one structural type to another. They range from nearly circular to elliptical to odd shapes such as tear drops (Wu and Landolt, 1986). Figure 2.2 shows the projections of the main channels for some of the zeolites.

Among the zeolites in this group, only ZSM-5 and ZSM-11 have bidirectional intersecting channels. The others have non-intersecting unidirectional channels.

ZSM-5 has received the most attention by far, and serves as a prime example of the uniqueness of medium pore zeolites in catalysis. ZSM-5 can be synthesized by including organic molecules such as tetrapropylammonium bromide as a template in the reaction mixtures. The organic molecules are incorporated into the zeolite crystals filling the intracrystalline void space either as organic cations or as occluded salt molecules (Dessau et al., 1987) when the zeolite is crystallized from solution (Argauer and Landolt, 1972; Rollmann, 1979b; Kerr, 1981). Other organic reagents which have been used as templates in synthesis include amines, diamines and alcohols.

Within a much narrower compositional space, ZSM-5 can also be synthesized in the absence of an organic template (Plank et al., 1979; Grose and Flanigan, 1981; Wang et al., 1981; Bezak and Mostowicz, 1985; Nastro, 1985).

HZSM-5 and HZSM-11 are remarkably stable as acidic catalysts. Unlike other zeolites, they have pores of uniform dimension and have no large supercages with smaller size windows. The absence of bottlenecks in their pore system is believed to be a significant factor for their unusually low coke forming propensity as acidic catalysts. Other contributing factors to their low coking tendencies include their high silica-to-alumina ratios and the geometrical constraint imposed by the 10-membered oxygen-ring-sized pores. These factors make it sterically difficult to form the large polynuclear hydrocarbons responsible

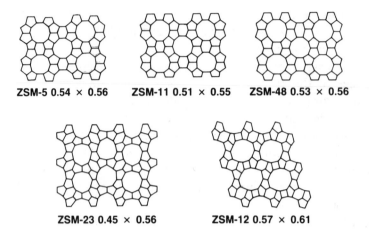

ZSM-5 0.54 × 0.56   ZSM-11 0.51 × 0.55   ZSM-48 0.53 × 0.56

ZSM-23 0.45 × 0.56   ZSM-12 0.57 × 0.61

Figure 2.2 Projections of ZSM-5, -11, -12, -23, and -48 structures. (From Wu et al., 1986.)

for coking and irreversible deactivation (Walsh and Rollmann, 1979; Rollmann and Walsh, 1979; Rollmann and Walsh, 1982; Derouane, 1985). This nonaging feature is probably one of the major reasons for the successful industrial application of these zeolites.

C. Dual Pore Systems

Zeolites in this group have interconnecting channels of either 12- and 8-membered oxygen ring openings or 10- and 8-membered oxygen ring openings. Examples are dachiardite, epistilbite, ferrierite (Winquist, 1976), gmelinite, heulandite/clinoptilolite, Linde T, mordenite, offretite, stilbite, ZSM-35 (Plank et al., 1977), etc.

As acidic catalysts, zeolites with intersecting channels of two different sizes, such as mordenite and offretite, also coke and deactivate rapidly. It is evident that the pore structure exerts a very large effect on the rate of coking. Large

12-membered oxygen ring openings or supercages deactivate much more rapidly than medium or small pores in acid cata-lyzed reactions (Venuto and Hamilton, 1967; Venuto, 1971; Rollmann and Walsh, 1977; Rollmann and Walsh, 1979; Dejaifve et al., 1981).

Some of these dual pore zeolites show interesting catalytic properties in cracking reactions. For example, cracking of paraf-fins over offretite (Chen et al., 1984) was found to yield more low-molecular-weight cracked products than that produced from medium pore zeolites. It was speculated that this is because of the availability of the smaller channels that are accessible only by the smaller molecules. However, most of the natural varieties and some of the synthetic samples contain numerous stacking faults, and in many catalytic reactions they behave like small pore zeolites (Miale et al., 1966; Kibby et al., 1974; Chen et al., 1978b).

III. STRUCTURAL FEATURES OF MEDIUM PORE ZEOLITES

As described earlier, the medium pore zeolites, unlike other zeolites, have pores of uniform dimension and have no large supercages with smaller size windows. The 10-membered oxygen ring opening channels in medium pore zeolites are intermediate in size as compared with the smaller 8-membered oxygen ring openings and the larger 12-membered oxygen ring openings present in such zeolites as zeolite X, zeolite Y, and mordenite.

The structure of ZSM-5 was reported by Mobil workers in 1978 (Kokotailo et al., 1978a). The zeolite has been synthesized over a range of silica-to-alumina ratios (Argauer and Landolt, 1972; Dwyer and Jenkins, 1976). Flanigan and her co-workers (1978a, 1978b) have the name "Silicalite" to a ZSM-5 having a high silica-to-alumina mole ratio, even though it has the same framework structure.

Highly siliceous zeolites may have ion exchange capacities in excess of their framework tetrahedral aluminum content

(Fegan and Lowe, 1984). These aluminum-independent ion exchange sites have been identified as framework siloxy anions, which serve as counter-ions for the organic cations, but they do not contribute to acid activity (Chester et al., 1985).

Of significance to shape selective catalysis is the presence of two intersecting channels formed by rings of 10 oxygen atoms. Figure 2.3 depicts a schematic view of such a channel system (Meisel et al., 1976). These two intersecting channels, both formed by 10-membered oxygen rings, are slightly different in their pore size (Meier and Olson, 1978). One sinusoidal channel, which has an elliptical opening (5.1 X 5.5 Å), runs parallel to the a-axis of the unit cell. The other channel, which is straight, runs parallel to the b-axis, and has a nearly circular (5.4 X 5.6 Å) opening. These two types of channels intersect to form a 3-dimensional network of pores.

ZSM-11, another medium pore zeolite, has two straight, elliptical channels intersecting at right angles. Both channels are 5.1 X 5.5 Å in pore size (Kokotailo et al., 1978b).

The other known medium pore zeolites, Theta-1 (ZSM-22), ZSM-23 and ZSM-48, have only unidirectional channels. It is interesting to note that although the size and shape of their pores are slightly different, the pores in Theta-1 (ZSM-22) (pore size 4.5 X 5.5 Å) and ZSM-48 (pore size 5.3 X 5.6 Å) are formed by the same set of 5- and 6-membered oxygen rings in the same order of arrangement as that forming the straight channels of ZSM-5 (pore size 5.4 X 5.6 Å), ZSM-11 (pore size 5.1 X 5.5 Å) and ferrierite. The order of arrangement is slightly different for the ZSM-23 pores which are tear drop shaped (pore size 4.5 X 5.6 Å) (Rohrman et al., 1985).

In addition to the use of X-ray and electron diffraction methods to determine the framework structures of these zeolites, many other complementary techniques have been applied to characterize their pore-channel system.

Sorption measurements using probe molecules of different critical dimensions have been used to gauge the effective size of these pores (Chen and Garwood, 1978; Olson et al., 1981). By

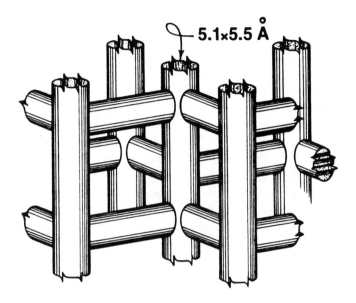

Figure 2.3   Channel system in ZSM-5. (From Meisel et al., 1976.)

the choice of sorbate molecules that have critical dimensions of different shapes (for example, 2,2-dimethyl butane has a nearly circular cross section; benzene has an elliptical cross section), the shape of the pore, circular or elliptical, may be differentiated (Wu and Landolt, 1986; Chester et al., 1987).

Catalytic diagnostic tests using model compounds of (Chen et al., 1978b; Dewing, 1984) different sizes and shapes have also become useful tools. For example, the relative rate of cracking of n-hexane and 3-methylpentane was used to differentiate members of the erionite/offretite family (Chen et al., 1984), and the clinoptilolite/heulandite family of zeolites (Chen et al., 1978b).

A similar test known as the "Constraint Index" test (Frilette et al., 1981) has been devised to distinguish medium pore zeolites from large pore and small pore zeolites.

Hydroisomerization of long chain paraffins, C_9 to C_{16}, was studied extensively by Jacobs, Weitkamp, and their co-workers (Jacobs et al., 1982; Weitkamp et al., 1983). Compared to large pore zeolites, such as the ultrastable Y (Jacobs et al., 1980), the channel structure of ZSM-5 was found to exert a pronounced effect on both the relative rate of cracking to isomerization and on the isomer distribution (Jacobs et al., 1980; Jacobs et al., 1982).

Jacobs and Teilen (1984) found that the shape selective effects of medium pore zeolites can be estimated from the product distribution of cyclodecane conversion. Dewing (1984) used the ratio of p-xylene to o-xylene in m-xylene isomerization to characterize the pore size of various medium pore zeolites.

With the advent of high speed computers and sophisticated computer graphics, it is conceivable that molecular mechanics calculations may soon be used to provide a realistic picture of the motion of guest molecules in a zeolite (Reischman and Olson, 1986).

IV. THE QUANTITATION OF ACIDIC FRAMEWORK ALUMINUM SITES IN MEDIUM–PORE HYDROGEN ZEOLITES

That acid catalyzed shape selective reactions can be achieved over small pore zeolites with pore openings comprising 8-membered oxygen rings (~5 Å in diameter) demonstrates that the acid activity must have originated within the intracrystalline cavities. Over the years, a great deal of research has been devoted to elucidate the nature and the quantitation of acid sites in zeolites (see, for example, Beagley et al., 1984). In more recent years, a variety of physical characterization methods, including FT-IR spectroscopy, MAS-NMR spectroscopy, and thermogravimetric techniques such as temperature programmed desorption (TPD) in addition to acid-base titration has been developed. For example, temperature programmed desorption analysis was employed by Mikovsky and his co-workers to show that NH_4Y samples of different SiO_2/Al_2O_3 ratios (Mikovsky and Marshall, 1976; Mikovsky and Marshall, 1977; Mikovsky et al., 1979) contain acid sites of four different strengths that appear to increase as the number of aluminum atoms neighboring the protonic site decreases. Figure 2.4 shows a typical temperature-programmed ammonia desorption (TPAD) curve of an NH_4Y, where the symbols n_0, n_1, n_2, and n_3 refer to the number of nearest neighboring aluminum atoms associated with the acid site.

More recently, MAS-NMR spectroscopy has become a powerful tool for the identification of the nature of silicon, aluminum, and other atoms in zeolites. Interestingly, ^{29}Si-NMR studies on zeolite X and Y samples (Engelhardt et al., 1981; Ramdas et al., 1981; Klinowski et al., 1982) show five types of silicon sites corresponding to the number of nearest aluminum neighbors ranging from 0 to 4. Their population distribution changes with the SiO_2/Al_2O_3 ratio in a manner consistent with the result obtained by TPD analysis. Figure 2.5 shows some typical ^{29}Si-NMR spectra on zeolite NaX and NaY (Engelhardt et al., 1981). Use of probe molecules, such as trimethyl phosphine in combination with MAS-NMR to identify the nature of acid sites in zeolite Y was reported recently by Lunsford (1986). In addition to identifying four types of Brønsted acid sites, he found evidence for the development of a multiplicity of Lewis acid sites upon dehydroxylating a H-Y zeolite.

With low SiO_2/Al_2O_3 ratio large pore zeolites, the task of correlating various acid sites with their catalytic properties is made most difficult not only by the multiplicity of acid sites, but also by added complexities such as rapid coking, and heat and mass transport effects due to their high acid site density and their pore structure. Obviously this is an important task, and numerous studies have dealt with this subject. However, the state of our knowledge relating catalytic properties to other characterizing information, particularly with respect to the relevance of fundamental understanding to practical applications,

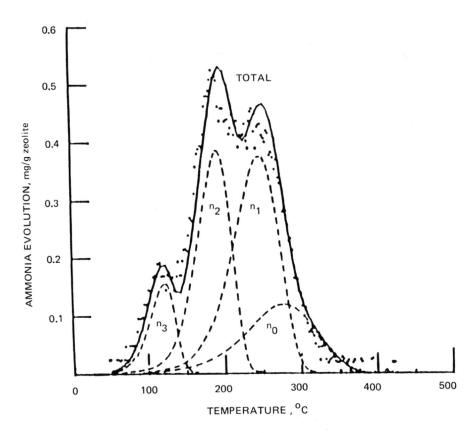

Figure 2.4   Typical temperature-programmed ammonia desorption (TPAD) curve of an NH_4Y. (From Mikovsky, 1976.)

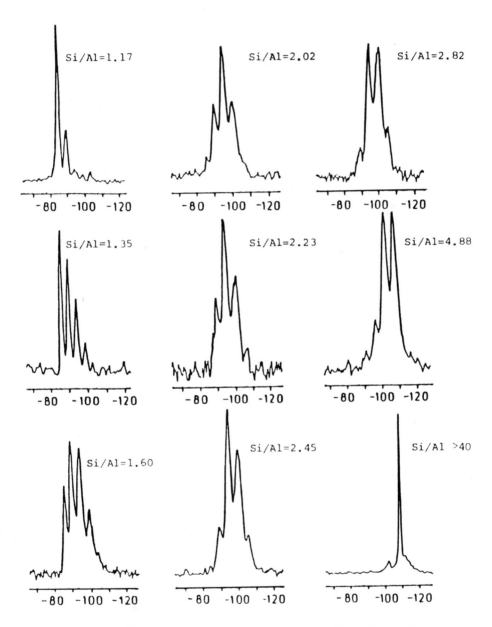

Figure 2.5 ^{29}Si NMR spectra of NaX and NaY zeolites. (From Englehardt et al., 1981.)

remains inadequate (see for example, Barthomeuf and Beaumont, 1973, Barthomeuf, 1979, Barthomeuf, 1980; Jacobs, 1982; DeCanio et al., 1986; Sohn et al., 1986).

The quantitation of protonic sites in medium pore zeolites, which generally have much higher SiO_2/Al_2O_3 ratios than the large pore zeolites, appears at first glance to be less complicated.

Using n-hexane (the α-test)* and n-hexene as the probing molecules and HZSM-5 samples carefully deammoniated from NH_4ZSM-5 in the absence of water vapor, an essentially linear relationship between catalytic activity and the number of protonic sites in HZSM-5 was established over a range of silica-to-alumina ratios, from 35 to 60,000 ($\sim$15 ppm Al). As shown in Figure 2.6, this linear relationship also extrapolates to zero activity at zero aluminum content (Olson et al., 1980). Similar relationships have been reported for a number of acid catalyzed reactions including the disproportionation of toluene, the conversion of methanol to hydrocarbons (Haag, 1984), ethylbenzene dealkylation, and cyclopropane isomerization (Chu et al., 1985). As shown in Figure 2.7, excellent linear relationships have also been found to exist between the elemental analysis, temperature programmed ammonia desorption (TPAD), and Al MAS-NMR (Haag et al., 1984; Derouane et al., 1985). Results of using other techniques, including infrared analysis (Topsoe et al., 1981; Jacobs and Von Ballmoos, 1982), ion exchange capacity,

---

*The α test measures the intrinsic acid activity of a given sample on a relative basis by using n-hexane as the probe molecule.

The "α value" of a catalyst is defined as the ratio of the first order rate constant of the sample to that of an arbitrary standard, the 46 AI (activity index) amorphous silica-alumina catalyst measured at $538°C$. Thus, a sample has an α value of one when its activity at $538°C$ is the same as that of the standard. For most samples, measurements made at other than $538°C$ can be extrapolated to $538°C$ using a temperature correction factor corresponding to an apparent activation energy of 30 kcal/mol. However, for some small pore and medium pore zeolites, such as erionite and ZSM-5, which have different apparent activation energies, caution must be taken in making proper extrapolations.

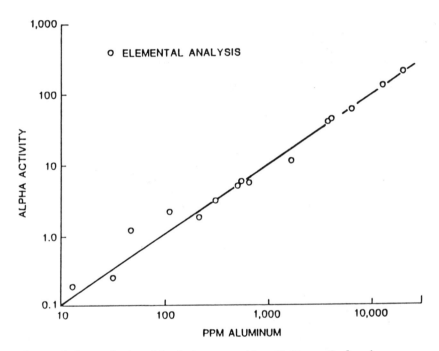

Figure 2.6   Relationship between acid activity and aluminum content. (From Olson et al., 1980.)

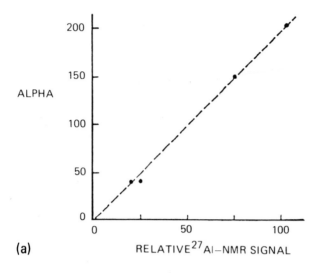

(a)

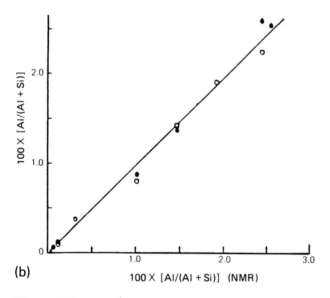

(b)

Figure 2.7 Relationships between catalytic activity, elemental analysis, and Al MAS-NMR. (a) Catalytic activity (alpha) vs. relative ^{27}Al-NMR signal. (b) Atomic fraction of framework Al from elemental analysis (●) vs. same from ^{27}Al MAS-NMR (○). (From Haag et al., 1984; Derouane et al., 1985.)

and reconstitution of NH_4^+ zeolite with ammonia (Beaumont and Barthomeuf, 1972; Jacobs, 1982), are consistent with the conclusion based on catalytic measurements that each framework Al-atom associated with a proton gives rise to one Brønsted acid site.

V. DISTRIBUTION AND LOCATION OF ACID SITES

Medium pore zeolites may be synthesized with only a trace concentration of aluminum. There is much speculation and uncertainty with respect to whether the placement of aluminum in the silica framework occurs by a random or ordered process, or whether aluminum atoms are located only in certain specific sites. Since the concentration of aluminum on the average is no more than one in ten of the tetrahedral framework positions, or a maximum of about 10 percent of the 96 Si+Al tetrahedrals per unit cell, the uncertainty is difficult to resolve.

Jacobs and von Ballmoos (1982) suggested that the acid sites are located at the channel intersections. Structurally speaking, every tetrahedral site is part of the channel intersections, although not all the oxygen atoms are located in the intersections. Therefore, the specific location of the protonic sites could conceivably affect its catalytic activity. This would be particularly pronounced in the case of bulky molecules. Derouane and his co-workers (Derouane and Vedrine, 1980; Dejaifve et al., 1980) also speculated that the channel intersections are probably the loci of catalytic activity of ZSM-5. But so far, this remains a largely unresolved issue.

Von Ballmoos and Meier (1981) also reported the presence of a concentration gradient of aluminum within a crystal. Higher aluminum concentration was found near the surface of the crystal relative to its bulk concentration (Von Ballmoos and Meier, 1981; Gabelica et al., 1984b). While it is still debatable whether these concentration gradients are present in any size crystals (Suib et al., 1980; Derouane et al., 1981), they are easily detectable in large crystals. Various synthesis methods have been designed to eliminate such gradients by maintaining a constant aluminum concentration during synthesis (Chen et al., 1978a) or to accentuate this gradient to produce a silicon-rich outer shell or an aluminum-rich shell by changing the gel composition abruptly during synthesis (Rollmann, 1978; Rollmann, 1979a; Miller, 1983; Koetsier, 1985). Thus, it is possible to synthesize zeolites possessing different spatially distributed acid sites. Similar concentration gradients can be expected when the zeolite is chemically modified to increase or decrease its framework aluminum concentration.

At least in principle, the spatial distribution of aluminum in the zeolite could be an important factor governing its effectiveness as a catalyst, since intracrystalline diffusion plays a major role in many reactions. This could be even more important in determining the shape selectivity of medium pore zeolite catalysts, because the degree of selectivity achievable could depend not only on the relative rate of diffusion of the feed and product molecules but also on the size of the crystals and the radial distribution of acid sites (Wei, 1982).

The contribution of the external surface of a zeolite crystal to its overall catalytic activity has often been questioned. With crystals larger than 1 micron, their external surface area relative to their intracrystalline surface area is so small (<1%) that surface activity can be ignored. However, for small submicron crystals, external surface sites could be a significant fraction of the total surface area. If the external surface sites are catalytically either the same or more active than the intracrystalline sites, then the shape selectivity of a zeolite could conceivably be changed by these surface sites (Gilson and Derouane, 1984). One such speculation was advanced by Fraenkel et al. (1984), who proposed that acid sites located in the "half-cavities" on the external crystal surface of ZSM-5 could be active for a second type of shape selective reaction. These sites were speculated to be responsible for the formation of such molecules as o,m-xylene, 1,2,4,5-tetramethyl benzene (durene) and 2,6- or 2,7-dimethyl naphthalene. However, the data as presented, appears insufficient to support this proposal, because in the absence of information on diffusion coefficients and reaction rates, one cannot rule out the effect of diffusional constraints on product selectivity. Furthermore, the surface sites are probably less active than tetrahydrally coordinated internal sites.

VI. STRUCTURAL EFFECT ON ACID SITES

A. Constraint Index

As acid catalysts, the medium pore zeolites have exceptional shape selectivity and low coking tendencies (Chen and Garwood, 1978; Walsh and Rollmann, 1979). As mentioned previously, a useful diagnostic test, known as the "Constraint Index," can be used to characterize these medium pore zeolites (Frilette et al., 1981). The numerical value determined by this test is approximately the ratio of the cracking rate constants for n-hexane and 3-methylpentane. Medium pore zeolites usually have constraint indices in the approximate range of 1 to 12.

The observed values of the constraint index for medium pore zeolites are largely independent of the crystal size of the zeolite. Therefore, their selectivity cannot be attributed to diffusion constraints of the reactant molecules. It is attributed to the steric constraint of the reaction intermediates inside the zeolite pores— a new type of shape selectivity, known as "spatiospecificity or transition state selectivity" (Haag et al., 1982), which will be discussed in more detail in Chapter 3.

This interpretation is consistent with the classical mechanism of acid-catalyzed paraffin cracking involving bimolecular hydride transfer as the rate-determining step. However, for a number of years there was no adequate explanation for the observation that the constraint index of medium pore zeolites decreases with increasing temperatures, shown in Figure 2.8 (Chen and Garwood, 1978; Frilette et al., 1981).

In a recent study, Haag and his co-workers (Haag and Dessau, 1984) demonstrated that the restricted pore geometry of

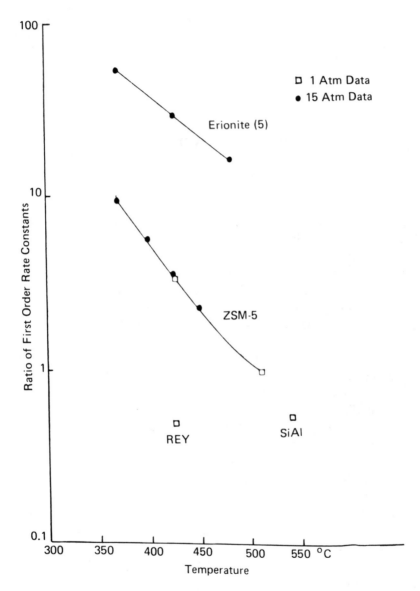

Figure 2.8 Shape selectivity between *n*-hexane and 3-methyl-pentane. (From Chen and Garwood, 1978.)

the medium pore zeolites promoted a second paraffin cracking mechanism involving the direct protonation of a paraffin molecule. The reaction is monomolecular and therefore does not involve a bulky reaction intermediate as does the classical mechanism. It is reasonable to expect that in the absence of steric constraint, n-hexane and 3-methylpentane should crack at comparable rates by this mechanism. The increasing importance of this reaction at higher temperatures provides a plausible explanation of this unusual temperature dependency of the constraint index of medium pore zeolites.

Because the cracked products are different for the reaction pathway of these two cracking mechanisms, it is possible to estimate from the product distirbution the contribution from each pathway.

B. Para-Selectivity

One of the unique shape selective characteristics of HZSM-5 is its para-selectivity (Chen et al., 1979) in electrophilic substitution reactions such as alkylation and disproportionation of alkyl aromatics.

By adjustment of the acid activity of the zeolite and controlling the diffusion parameter, high para-selectivity can be achieved. Generally, a necessary condition for good para-selectivity is that

$$1/k \ll R^2/D$$

where k is the reaction rate constant, R is the crystal radius, and D is the diffusivity of the species with the slowest diffusion rate.

Without changing the intrinsic activity of the catalyst and the size of the crystal, the effective diffusion characteristics of the catalyst can be altered by a number of techniques, including surface silylation (Chen et al., 1979); surface coking (Kaeding et al., 1984; Ashton et al., 1986); or surface poisoning with

bulky heterocyclic compounds (Chen, 1988); impregnation
with magnesium oxide (Kaeding et al., 1981a; Olson and Haag,
1984; Sato et al., 1985); antimony oxide (Butter, 1977);
phosphorus and boron compounds (Kaeding et al., 1981b; 1984).

Haag and his co-workers (Weisz, 1980; Plson and Haag,
1984) have developed a quantitative model relating the observed
para-selectivity with acid activity and diffusion parameters for a
large variety of modified and unmodified ZSM-5 catalysts.

VII. ENHANCED ACID SITES

While the characterization of protonic sites in the hydrogen
form of siliceous medium pore zeolites appears straightforward,
a general description of acid sites in zeolites is much more complex because hydrogen zeolites easily undergo thermal and hydrothermal reactions such as dehydroxylation, dealumination, and
stabilization. These reactions can alter the nature of the acid
sites (McDaniel and Maher, 1976; Poutsma, 1976).

It is well known that severe steaming of hydrogen zeolites
reduces the number of framework tetrahydral aluminum sites
and the catalytic acid activity (Chen and Smith, 1976). However, steaming at mild temperatures and under low partial pressure of water vapor and/or in the presence of ammonia can produce catalysts of higher activity than the parent zeolite. For
example, Haag and his co-workers (Lago et al., 1986) showed
that by steaming in the presence of ammonia, the catalytic
activity of a HZSM-5 was increased by a factor of 9. The
increase in catalytic activity can be explained by the presence of
a new type of acid site of higher specific activity. It is estimated
by the α test that the intrinsic specific activity of these enhanced
sites may be at least 40 times higher than that of the conventional protonic sites in ZSM-5. However, the nature of these
enhanced sites and the chemistry of their formation have not
been completely elucidated.

Equally interesting new ways of enhancing acidity by in-
serting aluminum into high-silica zeolite frameworks have
recently been reported in the literature.

Shihabi et al. (1985) showed that binding of high silica HZSM-5 with alumina enhances the catalytic activity of the catalyst for numerous reactions and postulated a mechanism for the insertion of aluminum from the binder into the tetrahedral framework of the zeolite.

Alumination of high-silica ZSM-5 has also been achieved with aluminum chloride (Anderson et al., 1984; Dessau and Kerr, 1984) and with a variety of aluminum halides or aqueous ammonium fluoroaluminate (Chang et al., 1984; Miale and Chang, 1984). At least some of the added aluminum is shown to be incorporated into the zeolite as a tetrahedrally coordinated species by independent physical measurements, such as ^{27}Al MAS NMR, NH_3-TPD, and FTIR. The treated catalysts show increased acid activity in methanol conversion and paraffin cracking reactions and give products similar to that of conventional HZSM-5 catalysts (Chang et al., 1984).

It remains unclear, however, whether these newly created acid sites have the same or different specific activity as the conventional protonic sites or the enhanced sites.

VIII. ACID SITES IN ISOMORPHOUS SUBSTITUTED ZEOLITES

The possibility of creating acid sites by isomorphous substitution of silicon with other elements, such as berylium, boron, chromium, gallium, germanium, iron, phosphorus, and titanium in addition to aluminum, by either direct synthesis or chemical modification has been receiving increased attention in recent years.

It is known that a number of natural minerals have the same framework topology but different chemical compositions (Barrer, 1982; Barrer, 1984). Early claims of making isomorphous substituted chromium zeolites by either the synthesis route (Ermolenko et al., 1964) or post treatments (Garwood et al., 1978) were made; however, positive identification of framework substitution was difficult. More recently, Taramasso et al. (1980) and Klotz (1981) reported that by X-ray analyses of the unit cell volume of borosilicates, a method was devised to determine the framework boron content of a number of boron substituted zeolites, including ZSM-5 and ZSM-11. The substitution of boron for silicon in ZSM-5 was further confirmed by ^{11}B MASNMR (Derouane et al., 1985). The insertion of boron was also accomplished using a Pyrex reaction vessel (Gabelica et al., 1984a) and by impregnation (Kaeding et al., 1981b; 1984; Gabelica et al., 1984a; Coudurier and Vedrine, 1986).

Available data on boron substituted zeolites (Taramasso et al., 1980; Holderich et al., 1984; Chu and Chang, 1985; Chang et al., 1985; Chu et al., 1985; Coudurier and Vedrine, 1986; Coudurier et al., 1987) indicates that the acid activity of the tetrahedrally coordinated framework boron sites is much lower than that of aluminum. In fact, Chu et al. (1985) attributed the observed catalytic acidity, if any, of B-ZSM-5 to trace amounts of aluminum sites present in the sample.

REFERENCES

Anderson, M. W., J. Klinowski, and X. Liu, J. Chem. Soc. Chem. Commun. 1596 (1984).

Argauer, R. J. and G. R. Landolt, U.S. Pat. 3,702,886, Nov. 14, 1972.

Ashton, A. G., S. Batmanian, J. Dwyer, I. S. Elliott, and F. R. Fitch, J. Mol. Catal. *34*, 73 (1986).

Barrer, R. M., Chem. Ind. (London), 1203 (1968).

Barrer, R. M., *Hydrothermal Chemistry of Zeolites*, Academic Press, London (1982).

Barrer, R. M., Proc. 6th Int. Zeol. Conf., D. H. Olson and Attilio Bisio, eds., Butterworths, Surrey, U.K., p. 870 (1984).

Barri, S. A. I., G. W. Smith, D. White, and D. Young, Nature *312*, 533 (1984).

Barthomeuf, D. and R. Beaumont, J. Catal. *30*, 288 (1973).

Barthomeuf, D., J. Phys. Chem. *83*, 249 (1979).

Barthomeuf, D., Stud. Surf. Sci. Catal. *5*, 55 (1980).

Beaumont, R. and D. Barthomeuf, J. Catal. *26*, 218 (1972).

Beagley, B., J. Dwyer, F. R. Fitch, R. Mann, and J. Walters, J. Phys. Chem. *88*, 1744 (1984).

Bennett, J. M. and J. A. Gard, Nature *214*, 1005 (1967).

Bezak, J. M. and R. Mostowicz, Stud. Surf. Sci. Catal. *24*, 47 (1985).

Breck, D. W., *Zeolite Molecular Sieves*, Wiley, New York (1974).

Butter, S. A. and W. W. Kaeding, U.S. Pat. 3,965,208, June 22, 1976.

Butter, S. A., U.S. Pat. 4,007,231, Feb. 8, 1977.

Chang, C. D., C. T.-W. Chu, J. N. Miale, R. F. Bridger, and R. B. Calvert, J. Am. Chem. Soc. *106*, 8143 (1984).

Chang, C. D., S. D. Hellring, J. N. Miale, and K. D. Schmitt, J. Chem. Soc., Faraday Trans. 1, *81*, 2215 (1985).

Chen, N. Y. and F. A. Smith, Inorg. Chem. *15*, 295 (1976).

Chen, N. Y., U.S. Pat. 4,002,697, Jan. 11, 1977.

Chen, N. Y., U.S. Pat. 4,100,215, July 11, 1978.

Chen, N. Y. and W. E. Garwood, J. Catal. *52*, 453 (1978).

Chen, N. Y., J. N. Miale, and W. J. Reagan, U.S. Pat. 4,112,056, Sept. 15, 1978a.

Chen, N. Y., W. J. Reagan, G. T. Kokotailo, and L. P. Childs, *Natural Zeolites*, L. B. Sand and F. A. Mumpton, eds., Pergamon, New York, p. 411 (1978b).

Chen, N. Y., W. W. Kaeding, and F. G. Dwyer, J. Am. Chem. Soc. *101*, 6783 (1979).

Chen, N. Y., J. L. Schlenker, W. E. Garwood, and G. T. Kokotailo, J. Catal. *86*, 24 (1984).

Chen, N. Y., J. Catal. *114*, 17 (1988).

Chester, A. W., Y. F. Chu, R. M. Dessau, G. T. Kerr, and C. T. Kresge, J. Chem. Soc., Chem. Commun. 289 (1985).

Chester, A. W., P. Chu, and W. J. Rohrbaugh, "Relationships between Measuring Zeolite Pore Sizes and Catalytic Performance," paper presented at the 10th North Am. Catal. Soc. Mtg., San Diego, May 17–22, 1987.

Chu, C. T.-W. and C. D. Chang, J. Phys. Chem. *89*, 1569 (1985).

Chu, C. T.-W., G. H. Kuehl, R. M. Lago, and C. D. Chang, J. Catal. *93*, 451 (1985).

Coudurier, G. and J. C. Vedrine, Pure and Appl. Chem. *58*, 1389 (1986).

Coudurier, G., A. Auroux, J. C. Vedrine, R. D. Farlee, L. Abrams, and R. D. Shannon, J. Catal. *108*, 1 (1987).

DeCanio, S. J., J. R. Sohn, P. O. Fritz, and J. H. Lunsford, J. Catal. *101*, 132 (1986).

Dejaifve, P., J. C. Vedrine, and E. G. Derouane, J. Catal. *63*, 331 (1980).

Dejaifve, P., A. Auroux, P. C. Gravelle, and J. C. Vedrine, J. Catal. *70*, 123 (1981).

Derouane, E. G. and J. C. Vedrine, J. Mol. Catal. *8*, 479 (1980).

Derouane, E. G., J. P. Gilson, Z. Gabelica, C. Mousty-Desbuquoit, and J. Verbist, J. Catal. *71*, 447 (1981).

Derouane, E. G., Stud. Surf. Sci. Catal. *20*, 221 (1985).

Derouane, E. G., L. Baltusis, R. M. Dessau, and K. D. Schmitt, Stud. Surf. Sci. Catal. *20*, 135 (1985).

Dessau, R. M. and G. T. Kerr, Zeolites *4*, 315 (1984).

Dessau, R. M., K. D. Schmitt, G. T. Kerr, G. L. Woolery, and L. B. Alemany, J. Catal. *104*, 484 (1987).

Dewing, J., J. Mol. Catal. *27*, 25 (1984).

Dwyer, F. G. and E. E. Jenkins, U.S. Pat. 3,941,871, Mar. 2, 1976.

Engelhardt, G., U. Lohse, E. Lippmaa, M. Tarmak, and M. Z. Magi, Anorg. Allg. Chem. *482*, 49 (1981).

Ermolenko, N. F., S. A. Levina, and L. V. Pansevitch-Kolada, Dokl. Akad, Nauk B. SSR *8*(6), 394 (1964).

Fegan, S. G. and B. M. Lowe, J. Chem. Soc., Chem. Commun. 437 (1984).

Flanigan, E. M., J. M. Bennett, R. W. Grose, J. P. Cohen, R. L. Patton, R. M. Kirchner, and J. V. Smith, Nature *272*, 840 (1978a).

Flanigan, E. M., J. M. Bennett, R. W. Grose, J. P. Cohen, R. L. Patton, R. M. Kirchner, and J. V. Smith, Nature *271*, 512 (1978b).

Fraenkel, R., M. Cherniavsky, and M. Levy, Proc. 8th Int. Congr. Catal. *4*, 545 (1984).

Frilette, V. J., W. O. Haag, and R. M. Lago, J. Catal. *67*, 218 (1981).

Gabelica, Z., G. Debras, and J. B. Nagy, Stud. Surf. Sci. Catal. *19*, 113 (1984a).

Gabelica, Z., E. G. Derouane, and N. Blom, Am. Chem. Soc. Symp. Ser. *248*, 219 (1984b).

Garwood, W. E., S. J. Lucki, N. Y. Chen, and J. C. Bailar, Jr., Inorg. Chem. *17*, 610 (1978).

Gilson, J. P. and E. G. Derouane, J. Catal. *88*, 538 (1984).

Gorring, R. L., J. Catal. *31*, 13 (1973).

Grose, R. W. and E. M. Flanigan, U.S. Pat. 4,257,885, Mar. 24, 1981.

Haag, W. O., R. M. Lago, and P. B. Weisz, J. Chem. Soc., Faraday Disc. *72*, 317 (1982).

Haag, W. O., Proc. 6th Int. Zeol. Conf., D. H. Olson and A. Bisio, eds., Butterworths, Surrey, U.K., p. 466 (1984).

Haag, W. O. and R. M. Dessau, Proc. 8th Int. Congr. Catal. *2*, 305 (1984).

Haag, W. O., R. M. Lago, and P. B. Weisz, Nature *309*, 589 (1984).

Holderich, W., H. Eichhorn, R. Lehnert, L. Marosi, W. Mross, R. Reinke, W. Ruppel, and H. Schlimper, Proc. 6th Int. Zeol. Conf., D. H. Olson and A. Bisio, eds., Butterworths, Surrey, U.K., p. 545 (1984).

Jacobs, P. A., J. B. Uytterhoeven, M. Steyns, G. Froment, and J. Weitkamp, Proc. 5th Int. Zeol. Conf., L. V. Rees, ed., Heyden, London, p. 607 (1980).

Jacobs, P. A., Catal. Rev.-Sci. Eng. *24*, 415 (1982).

Jacobs, P. A., J. A. Martens, J. Weitkamp, and H. K. Beyer, J. Chem. Soc., Faraday Disc. *72*, 353 (1982).

Jacobs, P. A. and R. Von Ballmoos, J. Phys. Chem. *86*, 3050 (1982).

Jacobs, P. A. and M. Teilen, Proc. 8th Int. Congr. Catal. *4*, 357 (1984).

Kaeding, W. W., C. Chu, L. B. Young, B. Weinstein, and S. A. Butter, J. Catal. *67*, 159 (1981a).

Kaeding, W. W., C. Chu, L. B. Young, and S. A. Butter, J. Catal. *69*, 392 (1981b).

Kaeding, W. W., L. B. Young, and C. C. Chu, J. Catal. *89*, 267 (1984).

Kerr, G. T., Catal. Rev.-Sci. Eng. *23*, 281 (1981).

Kibby, C. L., A. J. Perrotta, and F. E. Massoth, J. Catal. *35*, 256 (1974).

Klinowski, J., S. Ramdas, J. M. Thomas, C. A. Fyfe, and J. S.
    Hartman, J. Chem. Soc., Faraday Trans. 2 78, 1025 (1982).
Klotz, M. R., U.S. Pat. 4,268,420, May 19, 1981.
Koetsier, W., European Pat. EP 0 054 385 B1, Aug. 14, 1985;
    EP 0 055 044 B1, Sept. 18, 1985.
Kokotailo, G. T., S. L. Lawton, D. H. Olson, and W. M. Meier,
    Nature 272, 437 (1978a).
Kokotailo, G. T., P. Chu, S. L. Lawton, and W. M. Meier,
    Nature 275, 119 (1978b).
Kokotailo, G. T. and W. M. Meier, Chem. Soc., Spec. Publ. 33,
    133 (1980).
Kokotailo, G. T., J. L. Schlenker, F. G. Dwyer, and E. W.
    Valyocsik, Zeolites 5, 349 (1985).
Lago, R. M., W. O. Haag, R. J. Mikovsky, D. H. Olson, S. D.
    Hellring, K. D. Schmitt, and G. T. Kerr, Proc. 7th Int.
    Zeol. Conf., Y. Murakami, A. Iijima, and J. W. Ward, eds.,
    Kodansha/Elsevier, Tokyo/Amsterdam, p. 677 (1986).
Lok, B. M., C. A. Messina, R. L. Patton, R. T. Gajek, T. R.
    Cannan, and E. M. Flanigan, U.S. Pat. 4,440,871, Apr. 3,
    1984.
Lok, B. M., B. K. Marcus, E. M. Flanigan, U.S. Pat. 4,500,651,
    Feb. 19, 1985.
Lunsford, J. H., "Acid Sites in Zeolite Y: NMR Studies Using
    Probe Molecules," paper presented at the AIChE Nat. Mtg.,
    New Orleans, Apr. 6, 1986.
McDaniel, C. V. and P. K. Maher, Zeolite Chemistry and
    Catalysis, J. A. Rabo, ed., Am. Chem. Soc. Monogr. 171,
    316 (1976).
Meier, W. M. and D. H. Olson, Atlas of Zeolite Structure Types,
    Int. Zeolite Assoc., Polycrystal Book Service, Pittsburgh,
    1978.
Meier, W. M., Z. Kristallogr. 115, 439 (1979).
Meisel, S. L., J. P. McCullough, C. H. Lechthaler, and P. B.
    Weisz, Chemtech 6, 86 (1976).
Miale, J. N., N. Y. Chen, and P. B. Weisz, J. Catal. 6, 278 (1966).
Miale, J. N. and C. D. Chang, U.S. Pat. 4,427,786 through
    4,427,790, Jan. 24, 1984.
Mikovsky, R. J. and J. F. Marshall, J. Catal. 44, 170 (1976).
Mikovsky, R. J. and J. F. Marshall, J. Catal. 49, 120 (1977).

Mikovsky, R. J., J. F. Marshall, and W. P. Burgess, J. Catal. *58*, 489 (1979).

Miller, S. J., U.S. Pat. 4,394,362, July 19, 1983.

Nastro, A., C. Colella, and R. Aiello, Stud. Surf. Sci. Catal. *24*, 39 (1985).

Olson, D. H., R. M. Lago, and W. O. Haag, J. Catal. *61*, 390 (1980).

Olson, D. H., G. T. Kokotailo, S. L. Lawton, and W. M. Meier, J. Phys. Chem. *85*, 2238 (1981).

Olson, D. H. and W. O. Haag, Am. Chem. Soc. Symp. Ser. *248*, 275 (1984).

Plank, C. J., E. J. Rosinski, and M. K. Rubin, U.S. Pat. 4,016,245, Apr. 5, 1977.

Plank, C. J., E. J. Rosinski, and M. K. Rubin, U.S. Pat. 4,175,114, Nov. 20, 1979.

Poutsma, M. L., "Zeolite Chemistry and Catalysis," J. A. Rabo, ed., Am. Chem. Soc. Monogr. *171*, 437 (1976).

Ramdas, S., J. M. Thomas, J. Klinowski, C. A. Fyfe, and J. S. Hartman, Nature *292*, 228 (1981).

Reischman, P. T. and D. H. Olson, "A Theoretical Study of Hydrocarbon Packing in Shape Selective Zeolites," paper presented at the Florida Conf. Catal., Palm Beach, Fl., Apr. 21, 1986.

Rohrman, A. C., Jr., R. B. La Pierre, J. L. Schlenker, J. D. Wood, E. W. Valyocsik, M. K. Rubin, J. B. Higgins, and W. J. Rohrbaugh, Zeolites *5*, 352 (1985).

Rollmann, L. D., J. Catal. *47*, 113 (1977).

Rollmann, L. D., U.S. Pat. 4,088,605, May 9, 1978.

Rollmann, L. D., U.S. Pat. 4,148,713, Apr. 10, 1979a.

Rollmann, L. D., Adv. Chem. Ser. *173*, 387 (1979b).

Rollmann, L. D. and D. E. Walsh, J. Catal. *56*, 139 (1979).

Rollmann, L. D. and D. E. Walsh, "Progress in Catalyst Deactivation," Nijhoff, The Hague, p. 81 (1982).

Sato, H., N. Ishii, and S. Nakamura, U.S. Pat. 4,499,321, Feb. 12, 1985.

Schlenker, J. L., W. J. Rohrbaugh, P. Chu, E. W. Valyocsik, and G. T. Kokotailo, Zeolites *5*, 355 (1985).

Shihabi, D. S., W. E. Garwood, P. Chu, J. N. Miale, R. M. Lago, C. T. W. Chu, and C. D. Chang, J. Catal. *93*, 471 (1985).

Sohn, J. R., S. J. DeCanio, and J. H. Lunsford, J. Phys. Chem. 90, 4847 (1986).

Suib, S. L., G. D. Stucky, and R. J. Blattner, J. Catal., 65, 174 (1980).

Staples, L. W. and J. A. Gard, Min. Mag. 32, 261 (1959).

Taramasso, M., G. Perego, and B. Notari, Proc. 5th Int. Zeol. Conf., L. V. Rees, ed., Heyden, London, p. 40 (1980).

Topsoe, N. Y., K. Pedersen, and E. G. Derouane, J. Catal. 70, 41 (1981).

Venuto, P. B., L. A. Hamilton, and P. S. Landis, J. Catal. 5, 484 (1966).

Venuto, P. B., and L. A. Hamilton, Ind. Chem. Eng. Prod. Res. Dev. 6, 190 (1967).

Venuto, P. B. and P. S. Landis, Advan. Catal. 18, 259 (1968).

Venuto, P. B., Advan. Chem. Ser. 102, 260 (1971).

Von Ballmoos, R. and W. M. Meier, Nature 289, 782 (1981).

Von Ballmoos, R., Collection of Simulated XRD Powder Patterns for Zeolites, Butterworths, Surrey, U.K. (1984).

Walsh, D. E. and L. D. Rollmann, J. Catal. 49, 369 (1977).

Walsh, D. E. and L. D. Rollmann, J. Catal. 56, 195 (1979).

Wang, F., W. Cheng, and S. Chang, J. Catal. (Chinese) 2, 282 (1981).

Wei, J., J. Catal. 76, 433 (1982).

Weisz, P. B. and J. N. Miale, J. Catal. 4, 527 (1965).

Weisz, P. B., Proc. 7th Int. Congr. Catal. 1, 3 (1980).

Weitkamp, J., P. A. Jacobs, and J. A. Martens, Appl. Catal. 8, 123 (1983).

Wilson, S. T., B. M. Lok, and E. M. Flanigan, U.S. Pat. 4,310,440, Jan. 12, 1982a.

Wilson, S. T., B. M. Lok, C. A. Messina, T. R. Cannan, and E. M. Flanigan, J. Am. Chem. Soc. 104, 1146 (1982b).

Wilson, S. T. and E. M. Flanigan, U.S. Pat. 4,567,029, Jan. 28, 1986.

Winquist, B. H. C., U.S. Pat. 3,933,974, Jan. 20, 1976.

Wu, E. L., G. R. Landolt, and A. W. Chester, Proc. 7th Int. Zeol. Conf., Y. Murakami, A. Iijima and J. W. Ward, eds., Kodansha/Elsevier, Tokyo/Amsterdam, p. 547 (1986).

3
Principal Methods of Achieving Molecular Shape Selectivity

I. SIZE EXCLUSION

Separation among molecules of different sizes and shapes can be achieved by the proper choice of zeolites. The effective pore size of the zeolite can be further modified by the choice of cations associated with the zeolites. Catalytically, molecular shape selectivity by size exclusion can be achieved through either reactant selectivity or product selectivity. Reactant selectivity occurs when the feedstock contains two classes of molecules, one of which is too large to pass through the channel system of the zeolite. Product selectivity occurs when, among the multiplicity of products that could be formed, only those with the proper shape and size can pass through the channels as products.

Shown in the following tables are some of the early examples demonstrating the principle of size exclusion in shape selective catalysis. They include the use of 8-ring zeolites such as zeolite A, T, erionite, and chabazite, and stacking faulted 12-ring zeolites such as gmelinite and mordenite. The selective dehydration of n-butanol without reacting isobutanol was demonstrated over the zeolite CaA (Table 3.1) (Weisz et al., 1962). Reactant selectivity was also demonstrated by the selective cracking of n-hexane in the presence of 3-methylpentane (Table 3.2) (Miale et al., 1966). Product selectivity was demonstrated by near absence of branched chain products such as isobutane and isopentane (Table 3.3) (Weisz et al., 1962).

With medium pore zeolites, the cutoff point of molecular size exclusion is extended to include multiple branched aliphatic molecules and polyalkyl and polycyclic aromatics. However, for

Table 3.1   Dehydration of Primary Butyl Alcohols (1 atm.
Pressure with 6-sec Residence Time)

Temperature °C	130	230	260
Over Linde CaA molecular sieve			
isobutyl alcohol dehydrated, wt %		<2	<2
n-butyl alcohol dehydrated, wt %		18	60
sec-butyl alcohol dehydrated, wt %	~0		
Over Faujasite-type CaX molecular sieve			
isobutyl alcohol dehydrated, wt %		46	85
n-butyl alcohol dehydrated, wt %		9	64
sec-butyl alcohol dehydrated, wt %	82		

Source: Weisz et al. (1962).

many molecules, the cutoff depends on temperature (see Section III on configurational diffusion). For example, at room temperatures, 1,3,5-trimethylbenzene (mesitylene) does not enter the pores of ZSM-5 (Olson et al., 1981; Wu and Landolt, 1986). Chang (1983) showed that in the conversion of methanol to hydrocarbons over ZSM-5, where methylation of aromatic rings is an important reaction, hexamethylbenzene is not formed, and there is a sharp cutoff point in product molecular weight. Among the aromatics made from methanol (Table 3.4), 1,3,5-trimethylbenzene, 1,2,3,4- and 1,2,3,5-tetramethylbenzenes fall far short of their equilibrium concentrations.

Similarly, in the hydroisomerization of paraffins, Table 3.5, tribranched isomers are not formed and less amounts of dibranched isomers are formed when compared with the large pore zeolite Y (Jacobs et al., 1982).

Table 3.2 Molecular-Shape Selective Cracking

Catalyst	Hydrocarbon charge	Time on stream, min	Temperature, °C	Conversion, %
H Gmelinite	n-Hexane	10 to 33	370	47 to 30
H Gmelinite	2-Methylpentane	10 to 20	320 to 540	0 to 0.7
H Gmelinite	Methylcyclopentane	10 to 20	510 to 540	0.4 to 1.9
H Erionite	n-Hexane	26	320	52.1
H Erionite	2-Methylpentane	26	430	1.0
H Erionite	2-Methylpentane	26	540	4.7
H Chabazite	n-Hexane	30	260	10.0
H Chabazite	2-Methylpentane	10	540	1.5

Source: Miale et al. (1966).

Table 3.3   Comparison of n-Hexane and 3-Methylpentane
Cracking at 500°C

Catalyst	3-Methyl-pentane cracking conversion, %	n-Hexane Cracking		
		Conversion, %	Iso-C_4 / n-C_4	Iso-C_5 / n-C_5
96% silica chips	<1	1.1		
Amorphous silica-alumina (46AI)	28	12.2	1.4	10
Linde NaA	<1	1.4		
Linde CaA	<1	9.2	<0.05	<0.05

Source: Weisz et al. (1962).

Table 3.4   Aromatics Distribution from Methanol Conversion Over HZSM-5

	Normalized distribution, wt %	Normalized isomer distributions	Equilibrium distributions, 371°C
Benzene	4.1		
Toluene	25.6		
Ethylbenzene	1.9		
Xylenes:			
o	9.0	⌈ 21.5 ⌉	⌈ 23.8 ⌉
m	22.8	| 54.6 |	| 52.7 |
p	10.0	⌊ 23.9 ⌋	⌊ 23.5 ⌋
Trimethylbenzenes:			
123	0.9	⌈ 6.4 ⌉	⌈ 7.8 ⌉
124	11.1	| 78.7 |	| 66.0 |
135	2.1	⌊ 14.9 ⌋	⌊ 26.2 ⌋
Ethyltoluenes:			
o	0.7		
m + p	4.1		
Isopropylbenzene	0.2		
Tetramethylbenzenes:			
1234	0.4	⌈ 23.3 ⌉	⌈ 16.0 ⌉
1235	1.9	| 44.2 |	| 50.6 |
1245	2.0	⌊ 46.5 ⌋	⌊ 33.4 ⌋
Other A_{10}[a]	2.7		
A_{11}^{+}	0.4		

[a]Diethylbenzenes + dimethylethylbenzenes.
Source: Chang (1983).

Table 3.5   Distribution of the Feed Isomers from
n-Decane over Pt/H-Zeolites at 50% Total
Conversion

	Y*	ZSM-5	ZSM-11
Monobranched	65.0	95.0	85.0
Dibranched	30.0	5.0	15.0
Tribranched	5.0	—	—

Source: Jacobs et al. (1982).

II. COULOMBIC FIELD EFFECTS

The ionic character of the intracrystalline surface and the
strength of coulombic field interaction between the zeolite and
the sorbed molecules are functions of the silica-alumina ratio of
the zeolite and the exchanged cations. Sodium-exchanged low
SiO_2/Al_2O_3 zeolites have a strong electrostatic field while the
hydrogen form of high SiO_2/Al_2O_3 zeolites do not. Thus the
nature of the intracrystalline surface differs significantly among
these zeolites ranging from highly hydrophilic to substantially
hydrophobic. Chen (1976) showed that as framework aluminum
is removed from mordenite, the dealuminized zeolite becomes a
hydrophobic sorbent. The degree of hydrophobicity is dependent
on the SiO_2/Al_2O_3 ratio of the sample, suggesting that the adsorption of water involves a specific interaction with the tetrahedrally
coordinated aluminum and associated cation centers. Figure 3.1
shows a linear relationship between the water sorbed and the
concentration of alumina in the sample at a constant partial
pressure of water. The slope of the straight lines corresponds
to a coordination number of 4 water molecules per aluminum.
Olson et al. (1980) found a similar linear relationship between
the water sorption capacity of HZSM-5 and the alumina content
of the samples. These hydrophobic zeolites retain their hydrocarbon sorption capacity. This is shown in Figure 3.2 for mordenite and ZSM-5 of different SiO_2/Al_2O_3 ratios (Chen, 1976;

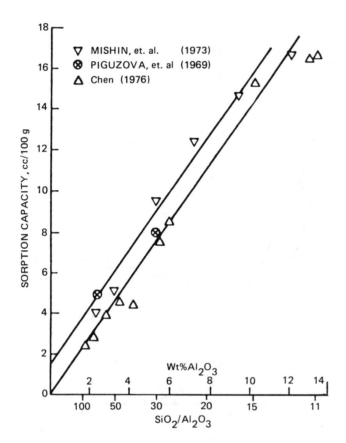

Figure 3.1 Water sorption at 25°C and 12 mmHg of water. (From Chen, 1976.)

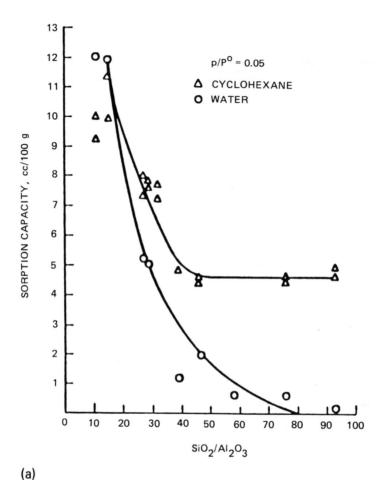

(a)

Figure 3.2   (a) Intracrystalline sorption at 25°C. (b) Amounts
of water, methanol, benzene, and n-hexane adsorbed on HZSM-5
at 100°C as a function of their SiO_2/Al_2O_3. (●) water, (△)
methanol, (□) n-hexane. (From (a) Chen, 1976, and (b) Nakamoto and Tokahashi, 1982.)

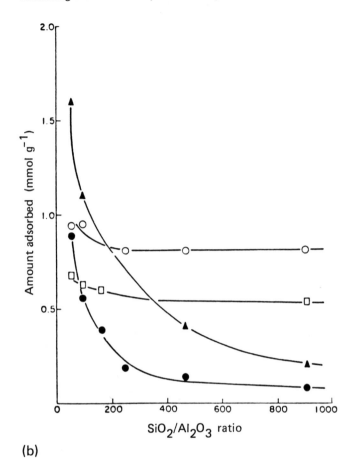

(b)

Figure 3.2   (Continued)

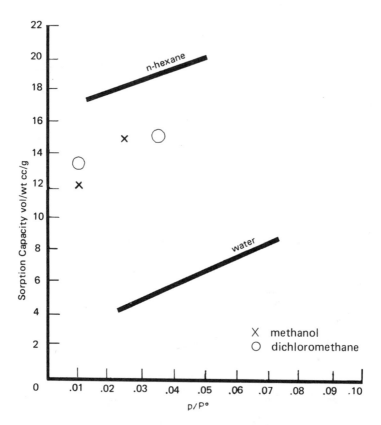

Figure 3.3 Adsorption isotherms at 25°C, ZSM-5 (SiO_2/Al_2O_3 = 54). (From Chen, 1973.)

Nakamoto and Takahashi, 1982). Shown in Figure 3.3 are the adsorption isotherms of water, *n*-hexane, methanol and dichloromethane at 25°C for a 54/1 SiO_2/Al_2O_3 HZSM-5 (Chen, 1973). With these hydrophobic zeolites, it is possible to separate polar compounds from less polar compounds, such as water and alcohols (Chen and Miale, 1983). Goldstein (1967) showed that acetic acid having a smaller van der Waal's radius than *cis*-butene-2, is completely excluded by the zeolite CaA while the larger *cis*-butene-2 molecule is sorbed. Thus a more sophisticated selectivity could be achieved beyond the realm of simple size exclusion.

III. CONFIGURATIONAL DIFFUSION

Configurational diffusion (Weisz, 1973) occurs in situations where the structural dimensions of the catalyst approach those of molecules. In this diffusion regime, even a subtle change in the dimensions of a molecule can result in a large change in its diffusivity, as shown in Figure 3.4.

For example, the diffusivity of *trans*-butene-2 is at least 200 times that of *cis*-butene-2 in zeolite CaA even though these two molecules differ in size by only about 0.2 Å (Chen and Weisz, 1967). Thus the rate of hydrogenation of *trans*-butene-2 over Pt/zeolite A can be much faster than that of *cis*-butene-2, as shown in Table 3.6.

To achieve molecular shape selectivity, not only the size but also the dynamics of the molecular structure needs to be considered. For example, a cage or window effect was found with zeolite T (Gorring, 1973) and erionite (Chen et al., 1969) that imposes a strong but varying diffusional constraint on the diffusing paraffinic molecules of different chain lengths. As shown in Figure 3.5, when a n-C_{22} paraffin or a n-C_{23} paraffin is cracked over erionite, there are two distinct maxima in the size distribution of the product molecules and a distinct minimum at a carbon number of 8. This nonlinear chain length effect was also observed in hydrocracking n-paraffins of varying chain length, Figure 3.6, over erionite (Chen and Garwood, 1973) and ferrierite (Giannetti and Perrotta, 1975). Interestingly, as shown by Figure 3.7, the diffusivity of n-paraffins in zeolite T, which has a channel system similar to erionite, also changes in a similar periodic pattern by over two orders of magnitude between the minimum at C_8 and the maxima at C_4 and C_{11}. Similar experiments with chabazite, which has a shorter unit cell than that of erionite, show a similar periodicity but the minimum n-paraffin diffusivity shifts down by about 2 carbon numbers, corresponding

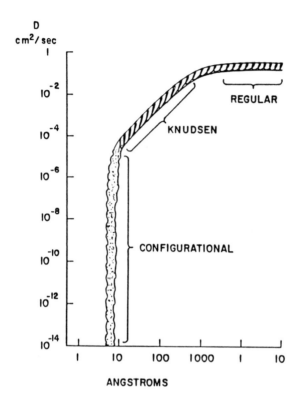

Figure 3.4   Configurational diffusion.   (From Weisz, 1973.)

to the unit cell size difference between these two zeolites, as shown in Figure 3.8 (Gorring and Daniels, 1987).

An example of this configurational diffusion effect on aromatics selectivity is the aforementioned achievement of para-directed aromatics alkylation and toluene disproportionation reactions (Chen et al., 1979). By increasing the crystallite size of ZSM-5, an increase in the diffusional constraints is imposed on the bulkier slower diffusing o- and m-isomers, reduces the production of these isomers, and increases the yield of the para-isomer, Table 3.7 (Olson and Haag, 1984).

Table 3.6 Hydrogenation of a Mixture of Trans- and Cis-Butene-2

Temp., °C	Initial composition		Final composition			Conversion ϵ, wt. %		$\dfrac{k_{trans}}{k_{cis}}$ [a]
	trans	cis	trans	cis	n-Butane	trans	cis	
120	78.7	21.3	37.1	17.0	45.9	52.9	10.8	3.3
103	78.7	21.3	57.3	19.8	22.9	27.2	7.1	4.3
98	78.7	21.3	69.4	20.9	9.7	11.8	1.8	7.0

[a] $k_{trans}/k_{cis} = \ln(1 - \epsilon_{cis})/\ln(1 - \epsilon_{trans})$.
Source: Chen and Weisz (1967).

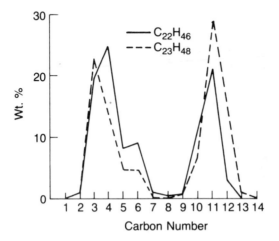

Figure 3.5   Carbon-number distribution of cracked products
over erionite at 340°C. (From Chen et al., 1969.)

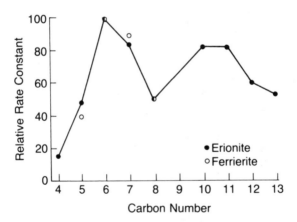

Figure 3.6   Hydrocracking aptterns over shape-selective zeolite
hydrocracking catalysts. (From Gianneti and Perrotta, 1975.)

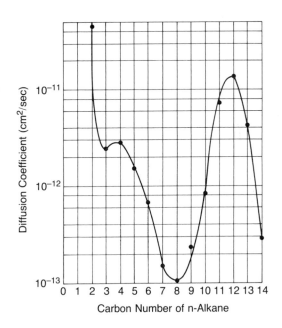

Figure 3.7   Diffusion coefficients of *n*-paraffins in potassium T at 300°C. (From Gorring, 1973.)

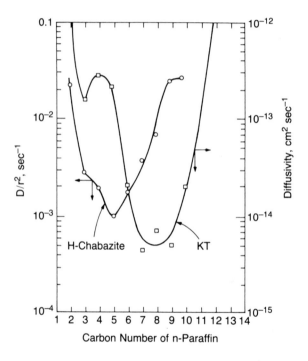

Figure 3.8   Diffusion of *n*-paraffins in H-chabazite KT. (From Gorring and Daniels, 1987.)

Table 3.7   Results with Large-Crystal ZSM-5

	Alkylation	Dispropor-tionation	Thermody-namic equilibrium
Temp,[a] °C	500	550	
WHSV[b]	6.6	30	
Feedstock	2:1 mol ratio of toluene/ methanol	toluene	
Conversion, wt %			
toluene	39	13.2	
methanol	99		
Product distri-bution, wt %			
C_1-C_5	2.6	<0.1	
benzene	1.9	5.5	
toluene	54.0	86.8	
xylenes			
para	17.9	2.6	
meta	14.0	3.5	
ortho	7.0	1.4	
others	3.3	0.1	
% Xylenes			
para	46	35	23
meta	36	46	51
ortho	18	19	26

[a] Weight hourly space velocity, (g of feed)/(g of catalyst) h^{-1}.
[b] Organic phase.
Source: Chen et al. (1979).

With the 10-ring ZSM-5 catalyst, Weisz et al. (1979) demonstrated that a triglyceride as large as $C_{57}H_{104}O_6$ can be converted
to the same product spectrum of hydrocarbon products as
methanol. This suggests that the triglyceride molecule can
easily attain, through molecular dynamics, a critical dimension
small enough to enter the channels of ZSM-5.

IV. SPATIOSPECIFICITY, OR TRANSITION STATE SELECTIVITY

When both the reactant molecule and the product molecule are
small enough to diffuse through the channels, but the reaction
intermediates are larger than either the reactants or the products
and are spatially constrained either by their size or by their
orientation, we term this *spatiospecific selectivity* to distinguish
it from stereospecificity. It is perhaps one of the most important
properties of ZSM-5, making it superior to other catalysts in a
number of important petrochemical processes.

Spatioselectivity or transition state selectivity is independent
of crystal size and activity, but depends on the pore diameter
and zeolite structure. This type of selectivity was first proposed
by Csicsery (1971) when he observed the absence of symmetrical
trialkylbenzenes in the product from the disproportionation of a
dialkylbenzene over H-mordenite. Since the reaction is bimolecular, the diphenylmethane type intermediates must require more
space than is available in the mordenite channels. Alternative
interpretations of the experimental results, on the basis of size
exclusion and configurational diffusion inhibition, were ruled
out by intramolecular isomerization experiments that indicated
that diffusion of symmetrical trialkylbenzenes in mordenite is
not hindered.

Spatiospecific selectivity plays a major role in the selective
cracking of paraffins in medium pore zeolites. For example,
n-hexane and 3-methylpentane are readily sorbed by ZSM-5, yet
the singly-branched molecule cracks at a significantly slower rate
than the straight chain molecule (Chen and Garwood, 1978).

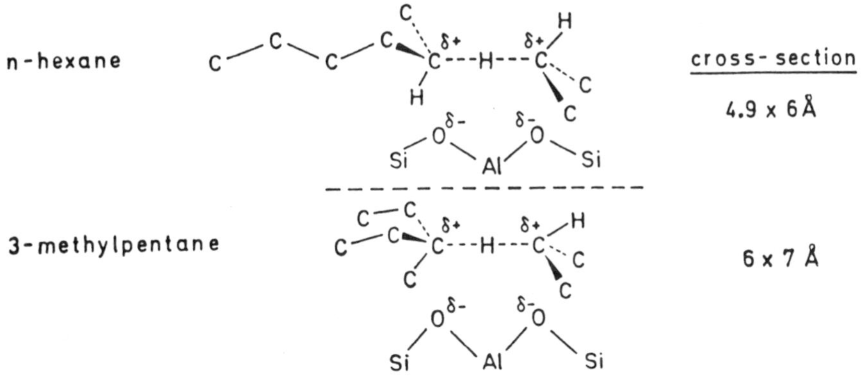

Figure 3.9   Mechanism of paraffin cracking. (From Haag et al., 1982.)

3-Methylpentane, being a bulkier molecule than *n*-hexane, apparently requires more space than *n*-hexane to form the reaction intermediate, as shown in Figure 3.9. Definitive studies in support of spatial constraints of reaction intermediates rather than configuration diffusion inhibition of reactants were made by Frilette, et al. (1981). They showed that the relative rate of cracking of *n*-hexane to that of 3-methylpentane is independent of the size of the individual catalyst crystal (Figure 3.10). Had diffusional constraint been an important factor in this reaction, the relative rate of cracking would have been affected by changing the crystallite size.

Amelse (1987) in studying the mechanism of transethylation over medium pore and large pore zeolites concluded that the mechanism for transethylation is different because of the spatioselectivity of the medium pore zeolites. The formation of a biphenylethane intermediate, which determines the rate of transethylation in large pore zeolites, is inhibited by the spatial constraints of the pores of the medium pore zeolites. Thus the transethylation of ethylbenzene over medium pore zeolites takes place via a different mechanism, namely, dealkylation and realkylation of the ethyl group.

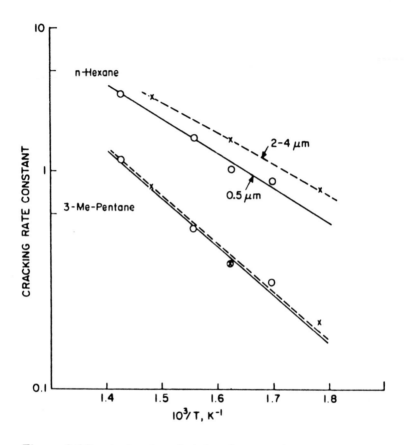

Figure 3.10   Arrhenius plot for the cracking of hexane isomers by HZSM-5 of different crystal size. (From Frilette et al., 1981.)

V. TRAFFIC CONTROL

A number of zeolites, including ZSM-5, ferrierite, clinoptilolite, offretite, and mordenite have intersecting channels of differing channel size (Table 2.1). Since the smaller channels are accessible only by the smaller molecules while the larger channels are accessible by both the large and small molecules, a new type of shape selectivity can be envisaged.

This concept was first proposed by Derouane, who termed it the molecular traffic control effect (Derouane and Gabelica, 1980). Derouane used this concept to explain the unusual absence of counter-diffusion effects in the ZSM-5 catalyzed conversion of simple molecules such as methanol. According to the concept of traffic control, the smaller molecules enter the sinusoidal channels while the larger product molecules exit from the elliptical channels. The significance of this effect was tested by comparing the catalytic properties of ZSM-5 with ZSM-11, the latter having intersecting straight channels of the same size (Derouane et al., 1982). But for such reactions as the conversion of methanol to hydrocarbons and the alkylation of toluene with methanol, their results show no major difference in the catalytic activity and stability of these two zeolites other than minor differences in product distribution, which thus largely negated the importance of this effect.

REFERENCES

Amelse, J. A., "A Shape Selective Shift in the Mechanism of Transalkylation and its Effect on the Ability to Hydrodeethylate Ethylbenzene," Paper B-4, 10th North Am. Mtg. Catal. Soc., San Diego, May 17–22, 1987.

Chang, C. D., Catal. Rev.-Sci. Eng. 25, 1 (1983); *Hydrocarbons from Methanol*, Marcel Dekker, New York, 1983.

Chen, N. Y. and P. B. Weisz, Chem. Eng. Prog. Symp. Ser. 73, 86 (1967).

Chen, N. Y., S. J. Lucki, and E. B. Mower, J. Catal. *13*, 329 (1969).

Chen, N. Y., U.S. Pat. 3,732,326, May 8, 1973.

Chen, N.Y. and W. E. Garwood, Advan. Chem. Ser. *121*, 575 (1973).

Chen, N.Y., J. Phys. Chem. *80*(1), 60 (1976).

Chen, N.Y. and W. E. Garwood, J. Catal. *52*, 453 (1978).

Chen, N. Y., W. W. Kaeding, and F. G. Dwyer, J. Am. Chem. Soc. *101*, 6783 (1979).

Chen, N. Y. and J. N. Miale, U.S. Pat. 4,420,561, Dec. 13, 1983.

Csicsery, S. M., J. Catal. *23*, 124 (1971).

Derouane, E. G. and Z. Gabelica, J. Catal. *65*, 486 (1980).

Derouane, E. G., P. Dejaifve, and Z. Gabelica, J. Chem. Soc., Faraday Disc. *72*, 331 (1982).

Frilette, V. J., W. O. Haag, and R. M. Lago, J. Catal. *67*, 218 (1981).

Gianneti, J. P. and A. J. Perrotta, Ind. Eng. Chem., Process Des. Develop. *14*, 86 (1975).

Goldstein, T. P., "The Silicon/Aluminum Ratio as a New Parameter in Determining Molecular Size Selective Sorption by Crystalline Aluminosilicates," paper presented at Am. Chem. Soc. 153rd Mtg., Miami Beach, Fl., Apr. 9–14, 1967.

Gorring, R. L., J. Catal. *31*, 13 (1973).

Gorring, R. L. and R. H. Daniels, unpublished data, 1987.

Jacobs, P. A., J. A. Martens, J. Weitkamp, and H. K. Beyer, Faraday Discuss. Chem. Soc. *72*, 353 (1982).

Miale, J. N., N. Y. Chen, and P. B. Weisz, J. Catal. *6*, 278 (1966).

Nakamoto, H., and H. Takahashi, Zeolites *2*, 67 (1982).

Olson, D. H., W. O. Haag, and R. M. Lago, J. Catal. *61*, 390 (1980).

Olson, D. H., C. T. Kokotailo, S. L. Lawton, and W. H. Meier, J. Phys. Chem. *85*, 2238 (1981).

Olson, D. H. and W. O. Haag, Am. Chem. Soc. Symp. Ser. *248*, 275 (1984).

Weisz, P. B., V. J. Frilette, R. W. Maatman, and E. B. Mower, J. Catal. *1*, 307 (1962).

Weisz, P. B., Chemtech *3*, 498 (1973).

Weisz, P. B., W. O. Haag, and P. G. Rodewald, Science *206*, 57 (1979).

Wu, E. L. and G. R. Landolt, "Pore Size and Shape Selective Effects in Zeolite Catalysis," paper presented at the 7th Int. Zeol. Conf., Tokyo, Aug. 17–22, 1986.

4
Shape Selective Acid Catalysis

I. REACTANTS OF INTEREST

It is generally accepted that hydrogen zeolites catalyze via carbenium ion intermediates, similar to reactions catalyzed by
strong acids in homogeneous media. Past reviews have examined
various reactions involving hydrocarbons, including paraffins,
olefins, naphthenes and aromatics (Venuto and Landis, 1968;
Poutsma, 1976).

With 8-membered-oxygen-ring small pore zeolites, shape
selective reactions deal only with the conversion of straight
chain molecules. With the availability of medium pore zeolites,
reactions involving nonstraight chain molecules and nonhydro-
carbon molecules are receiving increased attention. A recent
review by Chang (1983) examined the chemistry of hydrocarbons
from methanol and other O-compounds.

In the following sections, we will examine the impact of
shape selective catalysis on some of more common organic
reactions important to the petroleum and petrochemical industries. Among the reactants of interest are straight chain and
slightly branched chain paraffins and olefins; benzene and alkylbenzenes; single ring naphthenes and alkyl naphthenes; and non-
hydrocarbons, such as alcohols, ethers, acids, esters, pyridines,
phenols, and others.

II. REACTIONS CATALYZED BY MEDIUM
PORE ZEOLITES

A. Olefin Reactions

Olefins undergo a variety of acid catalyzed reactions, including
double bond isomerization, skeletal isomerization, oligomerization,

transmutation or disproportionation, cracking/polymerization, hydrogen transfer, and cyclization or aromatization.

With medium pore zeolites, these reactions are taking place under the constraints imposed by the size of the channels. As a result, unique products are formed as a function of reaction temperature and pressure.

The application of these olefin reactions over ZSM-5 to the production of gasoline and distillate from light olefins led to the development of Mobil's Olefin to Gasoline and Distillate Process (MOGD) (Tabak et al., 1986).

Isomerization

Olefins isomerize readily over acidic zeolite catalysts. For example, relative to the rate of cracking of n-hexane over HZSM-5, the rate of double bond shift of 1-hexene is at least 6 orders of magnitude faster (Haag et al., 1984), and is often diffusion controlled despite the fact that hexene is a relatively small molecule (Dessau, 1984). Consequently, the degree of branching of an olefin molecule with affect its rate of isomerization and molecular shape selectivity among the olefinic molecules is to be expected.

Isomerization—double bond shift

$$CH_3CH_2CH = CH_2 \rightleftharpoons CH_3-CH = CH-CH_3$$

$$CH_2 = \underset{\underset{CH_3}{|}}{C}-CH_2-CH_3 \rightleftharpoons CH_3-\underset{\underset{CH_3}{|}}{C} = CH-CH_3$$

Skeletal isomerization

$$CH_3-CH_2-CH_2-CH = CH_2 \rightleftharpoons CH_3-CH_2-\underset{\underset{CH_3}{|}}{C} = CH_2$$

Oligomerization

With medium pore zeolites, light olefins undergo rapid isomerization and oligomerization reactions at low temperatures with remarkable selectivity and stability not found with large pore zeolites.

$$2\ CH_3-CH_2-CH = CH_2 \ \rightleftharpoons \ CH_3-CH_2-\overset{\overset{\textstyle CH_3}{|}}{C}H-CH = CH-CH_2-CH_3$$

For example, Garwood (1983) showed that propene reacts to form predominantly trimers, tetramers and pentamers over HZSM-5 at 200°C, 2.7 WHSV and 35 atm (Figure 4.1). At longer residence times, olefin transmutation reactions convert the oligomers to secondary olefinic products. Similar results at atmospheric pressures were reported by Haag (Figure 4.2). More recently, the oligomerization activity of medium pore phosphate based molecular sieves, such as SAPO-11 and SAPO-31, was reported by Long et al. (1985) and Pellet et al. (1987).

Although no detailed data are available on the isomer distribution of the oligomers, it is expected that the oligomers produced by medium pore molecular sieve catalysts are less branched than those produced by nonshape selective acid catalysts (Haag et al., 1982a). Van Hooff and his co-workers showed that the oligmers formed in ZSM-5 at room temperatures from C_2 to C_4 olefins are linear in structure and proposed a mechanism involving stretching reactions via a cyclopropane intermediate to explain their formation (van den Berg et al., 1983). ^{13}C NMR analyses of propene oligomers also confirm qualitatively that the degree of branching is controlled by the dimensions of the zeolite cages and the pore structure (Galya et al., 1985).

Transmutation/Disproportionation

With medium pore zeolites such as HZSM-5, the secondary olefinic product from different feeds is substantially independent of the feed molecules. This suggests a rapid equilibration of olefins to a composition determined by the temperature and pressure of the system (Garwood, 1983).

The equilibration is generally perceived to be the result of a series of acid catalyzed polymerization and cracking reactions of

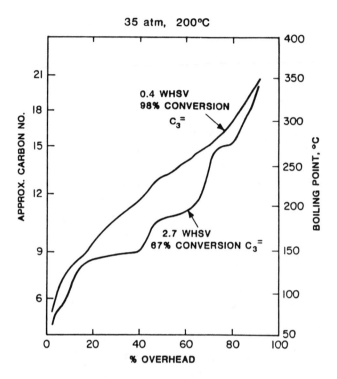

Figure 4.1   Effect of space velocity on propene conversion over HZSM-5.   (From Garwood, 1983.)

the oligomers, involving intramolecular rearrangement of carbenium ions, dimerization and β-scission.

Transmutation

$$C_4H_8 \rightleftharpoons \text{Mixture of } C_2 - C_{10} \text{ Olefins}$$

Disproportionation

$$
\begin{array}{c}
\text{CH}_3 \\
| \\
\text{CH}_3\text{-CH}_2\text{-C} = \text{CH-CH}_2\text{-CH}_2\text{-CH}_3 \rightleftharpoons \\
\\
\text{CH}_3 \\
| \\
\text{CH}_3\text{-CH}_2\text{-C} = \text{CH}_2 + \text{CH}_2 = \text{CH-CH}_3
\end{array}
$$

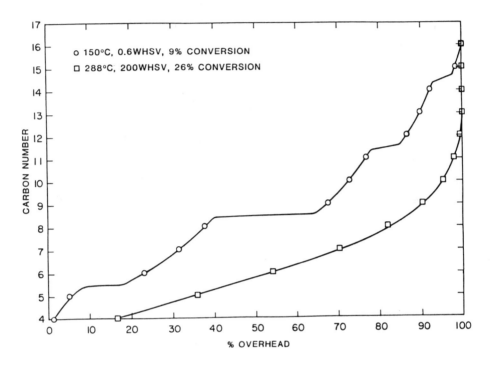

Figure 4.2   Effect of temperature and pressure on propene conversion. (From Haag, 1984.)

For example, Garwood (1983), showed that at $275°C$, propene, a mixture of pentenes, 1-hexene, and 1-decene all produced predominantly C_3–C_9 olefins (Table 4.1) over HZSM-5. They have a similar carbon number distribution and isomer distribution, which approaches chemical equilibrium. Table 4.2 compares the pentene product isomers with their equilibrium values.

The hexene isomer distribution is shown in Figure 4.3. When compared to the calculated equilibrium distribution, it is noted that even at as low as 3% conversion, the isomer distribution already approaches its equilibrium distribution. By following the extent of conversion, it can be seen that the reaction pathway begins with double bond shift followed by skeletal isomerization. The major deviation between the experimental and the calculated distributions are in the concentration of

Table 4.1   Olefin Equilibration Carbon Number Distribution

Product, wt %	Feed				
	Ethene	Propene	Pentene mix[b]	1-Hexene	1-Decene
Ethene	0[a]	<0.1	<0.1	<0.1	<0.1
Propene	11	8	10	9	4
Butenes	20	28	20	20	13
Pentenes	21	30	27	23	26
Hexenes	13	13	15	16	20
Heptenes	12	11	11	10	17
Octanes	8	6	7	8	8
Nonenes	8	3	5	6	7
C_{10}^{+} (unidentified)	7	1	5	8	5

[a] Based on converted ethenes.
[b] 88% 2-methyl-2-butene, 8% 2-methyl-1-butene.
Source: Garwood (1983).

2,3-dimethylbutenes, obviously the result of configurational diffusion constraints imposed by the zeolite.

Tabak et al. (1984; 1986) attempted to calculate the theoretical equilibrium composition in the range of temperature and pressure of interest. To avoid handling the astronomical number of isomers involved, the calculation was simplified by lumping each group of isomers of the same carbon number as a single compound and calculate the equilibrium among the lumped groups. To obtain the lumped free energy of a group requires the knowledge of the free energy of all the individual isomers, but this knowledge is at present available only up to 6 carbon atoms. Thus, for the higher-molecular-weight groups, an approximation technique developed by Alberty (1983) was used.

Table 4.2 Comparison of Pentene Product Isomers with Equilibrium at 275°C

	Feed					
	Ethene	Propene	Pentene mix	1-Hexene	1-Decene	Equilibrium
	Pentene isomer, %					
1-Pentene	2	2	2	2	2	2
2-Methyl-1-butene	18	16	18	18	17	24
3-Methyl-1-butene	1	2	2	2	1	2
trans-2-pentene	11	10	11	12	13	9
cis-2-pentene	5	5	5	5	6	7
2-Methyl-2-butene	63	65	62	61	61	56

Source: W. E. Garwood (1983).

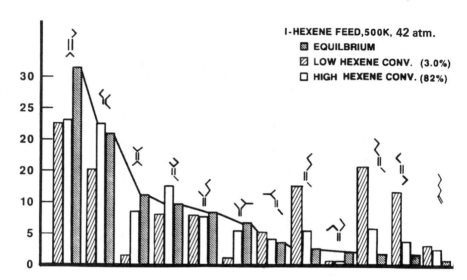

Figure 4.3   Comparison of experimental hexene isomer distribution with calculated equilibrium distribution. (From Quann et al., 1986.)

Because not all the isomers that theoretically exist can be produced by the ZSM-5, it was necessary to modify the parameters used in the extrapolation of free energies to obtain a reasonable agreement with the experimental results. Table 4.3 compares the theoretical values with the experimental data shown in Table 4.1, obtained at 275°C. The agreement is reasonably good.

At higher pressures, the product carbon number and boiling ranges increase with reaction temperature (Figure 4.4). For example, an olefinic mixture, consisting of 5 wt % butenes, 30 wt % pentenes, and 47 wt % hexenes (remaining 18% consisting of 12% C_4–C_6 paraffins, 6% C_7^+) was passed over HZSM-5 at 47 atm. As the reaction temperature was increased to 250°C, olefins boiling above 200°C became a significant part of the reaction product. Structural analysis of the 200°C$^+$ products

Table 4.3   Comparison of Calculated and Observed Olefin
Equilibria

Charge products, wt %	Propene		Pentene		Hexene	
	Data	Calc.	Data	Calc.	Data	Calc.
Ethene	<0.1	0.7	<0.1	0.5	<0.1	0.4
Propene	8	5.3	10	4.1	9	3.8
Butenes	28	23.9	20	20.2	20	19.1
Pentenes	30	21.2	27	19.6	23	19.1
Hexenes	13	23.2	15	23.6	16	23.6
Heptenes	11	12.5	11	13.9	10	14.3
Octanes	6	6.5	7	8.0	8	8.4
Nonenes	3	3.4	5	4.5	6	4.9
C_{10}^+ (unidentified)	1	3.5	5	5.6	8	6.4

Source: Tabak et al. (1986).

shows primarily methyl branching with one methyl branch for
each five carbons.

   At higher temperatures, the molecular weight distribution of
the product decreases. This is shown in Figure 4.5 (Quann et al.,
1986) for propene at 55 atm and 1 WHSV over the temperature
range of 204 to 382°C.

   Results of the theoretical equilibrium calculations made at
elevated pressures, when compared with the experimental results
as shown in Figure 4.6, show a large discrepancy at low temperatures. The calculation overestimates the average carbon number
of the product at below 330°C. Nevertheless, it shows the trend
of decreasing average carbon number at higher temperatures. The
calculated product distribution agrees reasonably with the experimental data at above 330°C.

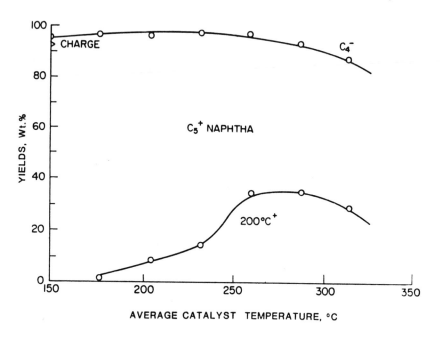

Figure 4.4 High boiling products from C_4, C_5, C_6 olefin mixture 1 LHSV, 47 atm. (From Garwood, 1983.)

Hydrogen Transfer/Cyclization

In addition to transmutation/disproportionation reaction, olefins undergo hydrogen transfer reactions to form cyclo-olefin. The formation of cyclo-olefins (C_nH_{2n-2}) from the reaction of 1-hexene over ZSM-5 is shown in Figure 4.7.

$$2 \text{ C--C--C--C} = \text{C--C} \rightleftharpoons \quad\bigcirc\!\!= \quad + \text{ RH}$$

The presence of cyclo-olefins was verified by hydrogenation of selected products and subsequent examination by FIMS (Quann et al., 1986).

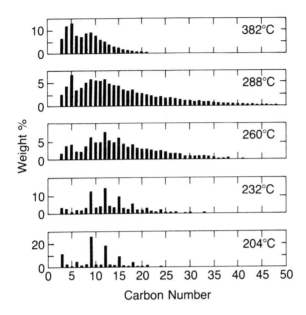

Figure 4.5   Propene polymerization at 55 atm, 1 WHSV. (From Quann et al., 1986.)

Hydrogen Transfer/Aromatization—M2-Forming

When the temperature is increased, a shift in product distribution toward BTX aromatics and light paraffins occurs (Table 4.4) as hydrogen transfer and cracking reactions become appreciable (Garwood, 1983).

Aromatization reactions of olefins become significant at about 370°C at atmospheric pressures over HZSM-5. M2-Forming, a generic name, has been coined to describe the reactions producing BTX aromatics from nonaromatic hydrocarbons over the medium pore zeolites (Chen and Yan, 1986).

The reaction products are again found to be independent of feed composition and are formed in a consecutive reaction pathway via cracking and hydrogen transfer reactions. The yield of aromatics is subject to the stoichiometric constraint of the carbon/hydrogen balance.

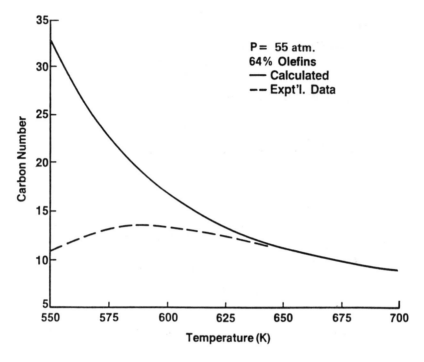

Figure 4.6   Average carbon number at equilibrium. (From Tabak et al., 1986.)

$$C_3^= \rightleftharpoons C_2\text{-}C_{10} \text{ Olefins} \xrightarrow{\text{Hydrogen Transfer}} B \cdot T \cdot X + C_2H_6 + CH_4 + H_2$$

Derouane and his co-worker observed that the channel inter-sections provide a large spherical space for two C_3-C_5 straight chain molecules to stack on top of each other and proposed a rationale for the formation of aromatics from aliphatics and the observed cutoff point of the aromatics that are made with the medium pore zeolites (Derouane and Vedrine, 1980; Dejaifve et al., 1980).

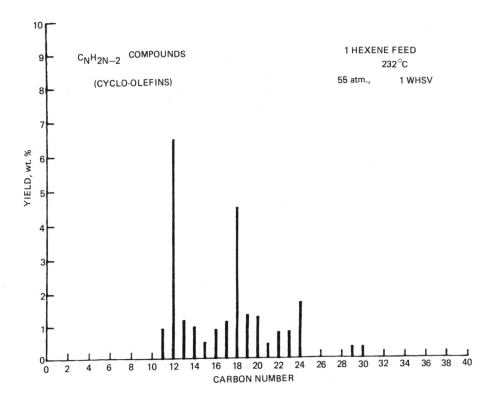

Figure 4.7   MOGD product distribution by FIMS. (From Quann et al., 1986.)

Ester Formation by Addition to Carboxylic Acids

The addition of an olefin to a carboxylic acid over acidic catalysts generally yields a mixture of esters. Young (1982) showed that α-methylalkyl carboxylates can be selectively produced over medium pore zeolites from carboxylic acids and olefins with either a terminal or an internal double bond.

Table 4.4   Conversion of Propene at 390°C, 1 atm, 0.5 WHSV

Carbon no.	Product, wt %				
	Olefin	Isoparaffin	n-Paraffin	Aromatics	Total
1	—	—	0.2	—	0.2
2	—	—	0.8	—	0.8
3	0.4	—	23.3	—	23.7
4	0.7	17.1	9.1	—	26.9
5	0.1	7.2	1.7	—	9.0
6	0.1	1.0	0.2	2.0	3.3
7	0.1	0.9	0.2	10.7	11.9
8	—	—	—	12.8	12.8
9	—	—	—	6.3	6.3
10	—	—	—	2.1	2.1
11^+	—	—	—	3.0	3.0
	1.4	26.2	35.5	36.9	100.0

Source: Garwood (1983).

$$R\text{-CH} = \text{CH-CH}_3 \rightleftharpoons R = \text{CH}_2 \rightleftharpoons R\text{-CH} = \text{CH-R}$$

$$\downarrow \quad + R\text{-C}\overset{O}{\underset{OH}{}}$$

$$R\text{-C}\overset{O}{\underset{\underset{\overset{|}{CH_3\text{-CH-R}}}{O}}{}}$$

Table 4.5 shows the result of adding 1-octene to acetic acid in a batch autoclave at 16–20 atm. It is interesting to note that, because of the fast double bond shift reaction and the size difference between the smaller α-methyl heptylacetate (2-octylacetate) and the other bulkier isomers and therefore the great difference in their relative rates of diffusion in medium pore

Table 4.5   Reaction of 1-Octene with Acetic Acid

Catalyst: ZSM-12, SiO_2/Al_2O_3 = 70

Acetic acid/1-octene — 4/1 molar

Reaction time, hr	T °C	P atm	Yield, wt %	$C_{18}H_{17}OAc$ isomer, wt %		
				2	3	4
3.5	150	12	5.9	96.9	2.9	0.2
5.6	200	17	23.9	94.1	5.5	0.4
73.2	200	17	32.2	91.4	8.1	0.5

Source: Young (1982).

zeolites, 2-octylacetate, a product of 1-octene and acetic acid, was selectively produced over ZSM-5 and ZSM-12. The selective production of α-methyl heptylacetate was even more dramatically demonstrated by using a mixture of octenes, consisting of 25% 1-octene, 25% *trans*-2-octene, 25% *trans*-3-octene and 25% *trans*-4-octene. As shown by the data in Table 4.6, instead of

Table 4.6   Reaction of Mixed Octenes with Acetic Acid

Feed: 25% 1-octene, 25% *trans*-2-octene,
25% *trans*-3-octene and 25% *trans*-4-octene

Reaction time, hr	T °C	P atm	Yield, wt %	$C_{18}H_{17}OAc$ isomer, wt %		
				2	3	4
1.75	150	12	14.1	88	11	1
2.75	150	12	17.0	86	13	1

Source: Young (1982).

getting an isomeric mixture of octylacetates, 2-octylacetate was selectively produced over ZSM-12.

In a separate study, Sato (1984) showed that when compared to zeolite Y and mordenite, HZSM-5 is a far better catalyst for the production of ethylacetate and isopropylacetate by adding to acetic acid ethene and propene respectively.

$$CH_3-C\overset{\displaystyle O}{\underset{\displaystyle OH}{\Big\langle}} \quad \begin{array}{l} + \; C_2H_4 \;\; \rightleftharpoons \;\; CH_3-C\overset{\displaystyle O}{\underset{\displaystyle OC_2H_5}{\Big\langle}} \\[2em] + \; C_3H_6 \;\; \rightleftharpoons \;\; CH_3-C\overset{\displaystyle O}{\underset{\displaystyle O-iC_3H_7}{\Big\langle}} \end{array}$$

Hydration

Chang and Morgan (1980) showed that hydration of C_2 to C_4 olefins to alcohols can be carried out over ZSM-5 at below about 240°C, and 10 to 20 atmospheres of pressure without forming ethers or other hydrocarbons.

$$C_3H_6 + H_2O \longrightarrow iC_3H_7OH$$

Above 240°C, however, propene and butenes undergo other olefinic reactions forming higher molecular weight hydrocarbon products. More recently, Chang and Hellring (1986) reported that glycols can be selectively synthesized by the hydrolysis of olefin oxides at 25°C over medium pore zeolites with excellent yields.

B. Paraffin Reactions

Isomerization

Hydroisomerization of light paraffins, such as C_4 to C_8 paraffins over medium pore zeolites has been patented by Mobil (Haag and Lago, 1983a). With enhancement of activity by steam and/or ammonia (Haag and Lago, 1983b), Pt/ZSM-5 was shown to be an excellent catalyst for the isomerization of n-pentane. The stability of the zeolite against coking allows the reactions to be carried out at lower hydrogen-to-hydrocarbons ratios and lower pressures.

Hydroisomerization of long chain paraffins, including C_9 to C_{16} paraffins was extensively studied by Jacobs, Weitkamp and their co-workers (Jacobs et al., 1982; Weitkamp et al., 1983). Based on detailed product distributions obtained from C_7-C_{16} paraffins, the following reaction network was proposed:

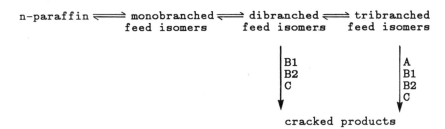

In this network the isomerization reaction proceeds via two routes: (1) type A isomerization (via alkyl shift), which is faster than (2) type B isomerication (via protonated cyclopropane). A, B1, B2 and C are four different types of β-scission reactions shown in Table 4.7.

Starting with normal and monobranched paraffins, isomerization clearly precedes hydrocracking reactions. Cracking of normal as well as of monobranched alkylcarbenium ions is a very slow process. Cracking starts only when the n-paraffin molecule has undergone at least two isomerization steps. Figure 4.8 shows that over a large pore Pt/ultrastable Y catalyst (Martens, 1985), high yields of isomerized product can be obtained before cracking becomes a significant factor. However, as the isomer assumes an α,α,γ-configuration, they are hydrocracked rapidly via the type A cracking route (tertiary to tertiary).

Compared to large pore zeolites, such as the ultrastable Y (Jacobs et al., 1980), the channel structure of ZSM-5 was found to exert a pronounced effect on both the relative rate of cracking

Table 4.7   Possible β-Scission Mechanisms on Secondary and Tertiary Carbocations

Type	Ions involved	Example
A	tert[a] → tert	
B1	sec[b] → tert	
B2	tert → sec	
C	sec → sec	

[a]tert = tertiary.
[b]sec = secondary.
Source: Weitkamp, et al. (1983).

to isomerization and on the isomer distribution. Thus, the hydroconversion of *n*-paraffins over Pt/ultrastable Y, Pt/ZSM-5 and Pt/ZSM-11 of comparable acidity and hydrogenation activity showed marked contrast in their product distribution (Jacobs et al., 1980; Jacobs et al., 1982).

   At 1 atm, with the ultrastable Y catalyst, decane isomerizes without cracking; with medium pore zeolites, the formation of bulky di- and tribranched decanes and ethyloctanes is suppressed. For example, at 5% conversion of decane, with Pt/ZSM-5 ethyloctanes are not produced and less than 10% of the isomers are dibranched.

   With an increase in pressure to 20 atm (Weitkamp et al., 1983), Pt/ZSM-5 can isomerize *n*-nonane without any hydrocracking, however *n*-pentadecane still cracks extensively and gives low yields of skeletal isomers, consisting primarily of mono-methyl paraffins.

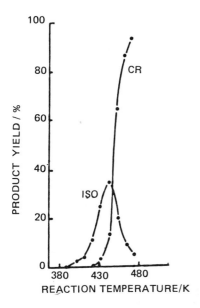

Figure 4.8 Yield of isomers (ISO) and cracked products (CR) from decane, converted over Pt/USY at various reaction temperatures. (From Martens, 1985.)

Cracking and Hydrocracking

The selective cracking of straight chain paraffins from a mixture of other hydrocarbons, was one of the initial demonstrations of shape selective catalysis. The first commercial shape selective catalytic process, the Selectoforming Process, uses erionite as the catalyst to selectively crack low octane n-paraffins in reformates (Chen et al., 1968).

Erionite is an effective cracking catalyst for gasoline range paraffinic molecules (C_{11} and lower) under moderate pressures, but requires much higher hydrogen pressures (above 68 atm) to crack higher boiling paraffins. For example, at 35 atm, there is a sharp break in the relative rate of conversion of n-paraffins to nonnormals (Chen and Garwood, 1978a). This pressure dependency is probably related to the diffusional characteristics of erionite (Gorring, 1973; Chen et al., 1968) but is not well understood.

Unlike erionite, the rate of cracking of paraffins over medium pore zeolite increases with the length of the molecules, and decreases with the bulkiness of the molecules. Figure 4.9 shows the relative rate of cracking of each component when a mixture of C_5 to C_7 paraffins was passed over HZSM-5 at 1.4 LHSV, 35 atm, and 340°C (Chen and Garwood, 1978b).

As mentioned in Chapter 3, spatiospecific selectivity and configurational diffusion effects influence dramatically the selective cracking of paraffins in the ZSM-5 family of zeolites. The former effect is particularly important in the relative rate of cracking of normal and monomethyl substituted paraffins, while the latter plays a major role in the cracking of multiple branched paraffins.

The extensive study on the mechanism of hydrocracking of long chain paraffins, by Jacobs, Weitkamp and their co-workers (Jacobs, 1982; Martens, 1985; Weitkamp et al., 1983) showed that in hydrocracking over Pt/Ultrastable Y, the relative rates of all six different types of carbenium ion reactions (isomerization of type A and B, and the four types of hydrocracking A, B1, B2 and C) are independent of chain length above C_{10}.

Hydrocracking of type A is the fastest reaction, followed in decreasing order by isomerization of type A, B1-cracking, B2-cracking, B-type isomerization and C-cracking, except that with increasing chain length, cracking of tribranched molecules via B1, B2 and C types is slightly faster than A-type, because the latter requires an α,α,γ-configuration, which needs several alkyl shifts to form. This explains the observed increased selectivity for di-branched cracked products in heavier fractions.

The primary cracked product of the four different types of β-scission reactions denoted with A, B1, B2 and C can be distinguished as they require specific configurations of the side chains. In the case of octane, nonane, and decane, 35%, 53% and 69% of the cracked products respectively are formed via the A type. The remaining products are formed via B1-, B2- and C-cracking.

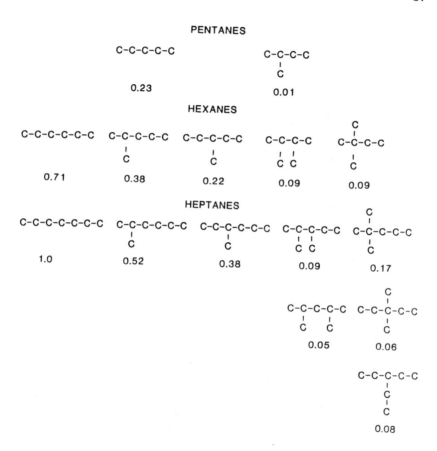

Figure 4.9   Relative rate of cracking of paraffins over ZSM-5
1.4 LHSV, 35 atm, 340°C. (From Chen and Garwood, 1978b.)

Detailed analyses of the composition of cracked products from the hydrocracking and hydroisomerization of n-paraffins over ZSM-5 (Weitkamp et al., 1983) suggest that reactions involving tertiary carbenium ion intermediates do not take place in medium pore zeolites. The relative high yield of C_3's, typical of medium pore and small zeolites, is the consequence of β-scission of secondary carbenium ions, or to a lesser extent, the not-favored primary carbenium ions.

The shape selective conversion of paraffins is the basis for a number of commercial petroleum refining processes, including M-Forming, a second generation post-reforming process (Heinemann, 1977) for upgrading the octane rating of reformates; MDDW, a distillate dewaxing process for increasing the quality and the yield of jet fuels, kerosene, and other distillate fuels (Chen et al., 1977; Donnelly and Green, 1980; Chen and Garwood, 1986) and MLDW, a catalytic lube dewaxing process, which replaces solvent dewaxing (Smith et al., 1980).

In addition to undergoing secondary cracking reactions, the olefinic cracked fragments also oligomerize, transmutate, cyclize, aromatize, and hydrogen-transfer, forming naphthenes, aromatics and lower molecular weight paraffins. Alkylation of the aromatics and a minor amount of ring closure to form dicyclics are also indicated. These can be seen in Table 4.8, which lists these products for n-octane conversion over ZSM-5 at 35 atm and 275°C.

Note also that the n-octane feed and its products are in almost quantitative hydrogen balance (Table 4.9). This balance explains the almost complete absence of coke on the catalyst, and the remarkable stability of the catalyst in the absence of a hydrogenation component and gaseous hydrogen. This is unique to the medium pore zeolites.

Aromatization—M2-Forming

At higher temperatures (up to 550°C) and atmospheric pressures, M2-Forming of paraffins to form predominantly BTX aromatics with hydrogen, methane, and ethane as the major by-products proceeds readily over medium pore zeolites.

Table 4.8 Products from n-Octane
Cracking 275°C, 35 atm, 1 LHSV

	Wt %
Methane	<0.1
Ethane	0.1
Propane	4.1
Propene	0.1
i-Butane	4.9
n-Butane	7.9
Butenes	0.1
i-Pentane	5.8
n-Pentane	9.6
C_6 Paraffins	4.8
C_7 Paraffins	2.4
C_8 Paraffins	49.7
Mono olefins	0.9
Diolefins	0.9
Benzene	1.1
Toluene	0.2
C_8 Alkylbenzenes	1.1
C_9 Alkylbenzenes	1.2
C_{10} Alkylbenzenes	0.6
C_{11} Alkylbenzenes	0.5
C_{12} Alkylbenzenes	0.4
C_{13} Alkylbenzenes	0.3
Tetralins, indanes	0.6
Naphthalenes	0.2
Monocyclic naphthenes	2.3
Bicyclic naphthenes	0.1
Mono olefins	0.9
Diolefins	0.9
	100.0

Source: Chen et al. (1979a).

Table 4.9 Hydrogen Balance

	n-Octane charge, 100 g	Products
Hydrogen content, g	15.89	15.85

Source: Chen et al. (1979a).

$$C_nH_{2n+2} \longrightarrow C_mH_{2m+2} + C_{n''}H_{2n''}$$

$$[C_2\text{-}C_{10} \text{ Olefins}]$$

$$(CH_2)_xC_6H_6 \qquad (BTX)$$

$$+$$

$$CH_4 + C_2H_6 + H_2$$

The yield of aromatics increases with decreasing hydrogen content of the feed, again subject to the stoichiometric constraints of carbon hydrogen balance (Chen and Yan, 1986).

C. Reactions of Aromatic Compounds

Alkylbenzenes undergo a variety of reactions over acid catalysts including isomerication, transalkylation/disproportionation, alkylation/dealkylation, and paring reactions (Sullivan et al., 1961). It is expected that the pore structure of medium pore zeolites would have a major influence on their stability and product distribution.

Isomerization

HZSM-5 shows an outstanding ability to isomerize xylenes with a minimum of side reactions, such as disproportionation to toluene and trimethylbenzenes. Haag and his co-workers (Haag and Dwyer, 1979; Olson and Haag, 1984) have shown that the observed high isomerization selectivity is probably the result of transition state selectivity, i.e., the bimolecular disproportionation of xylenes, which involves a bulky diphenylmethane type reaction intermediate that is sterically difficult to accommodate in the pores of the zeolite, is hindered relative to the monomolecular isomerization reaction, as shown in Figure 4.10 (Cortes and Corma, 1978; Haag et al., 1982a).

Based on these findings, several xylene isomerization processes have since been developed by Mobil.

It is also interesting to note that because of the large difference in the diffusion coefficient of p-xylene and o- and m-xylenes, $(D_p/D_{o,m}) > 10^4$, diffusional effects can give isomerized products containing p-xylene in concentrations exceeding that of its equilibrium value. An example of such a diffusion disguised kinetics is shown in Figure 4.11. In this case, the reaction pathway of o-xylene isomerization over the ZSM-5 catalyst was changed by treating the catalyst chemically to reduce its diffusion coefficient. It is noted that in the case of the chemically treated catalysts, the ratio of *para*- to *meta*-xylene exceeds its equilibrium value (Young et al., 1982). Similarly, high *para*- to *ortho*-xylene was observed when xylene was isomerized over a large crystal ZSM-5 (Ratnasamy et al., 1986).

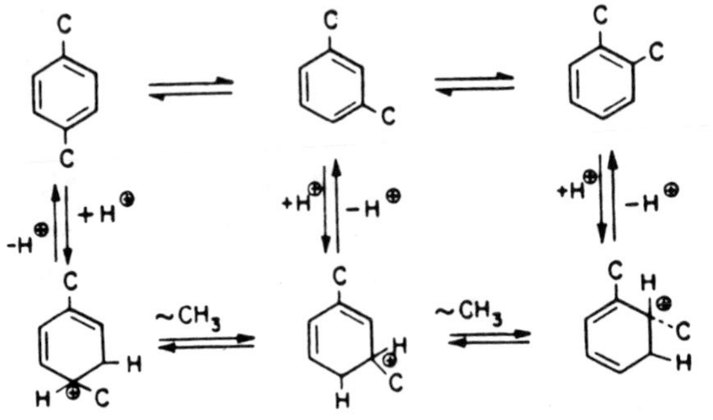

(a)

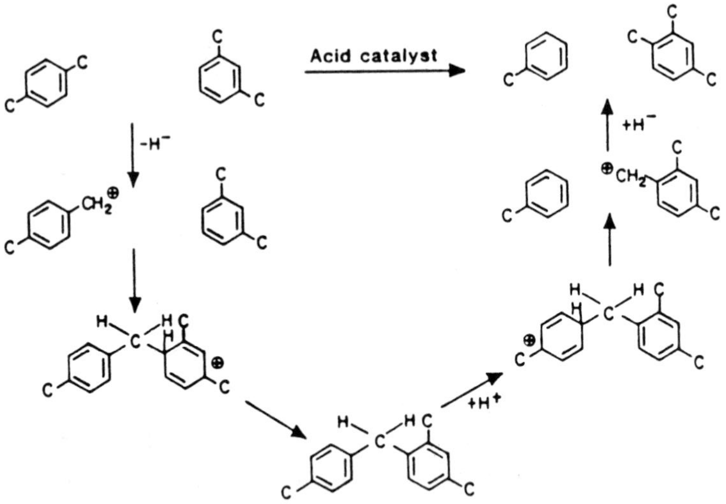

(b)

Figure 4.10 Comparison of monomolecular and biomolecular reactions of xylenes. (a) Acid-catalyzed xylene isomerization mechanism. (b) Acid-catalyzed xylene disproportionation mechanism. (From Olson and Haag, 1984.)

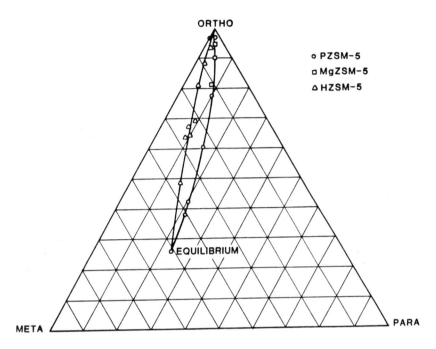

Figure 4.11   Reaction paths for isomerization of O-xylene.
(From Young et al., 1982.)

Transalkylation/Disproportionation

Because ethylbenzene is always present in commercial xylene
streams to the xylene isomerization process, the relative rate of
transalkylation of ethylbenzene to that of disproportionation of
xylene is an important factor in determining the selectivity of
the catalyst.

The transalkylation of ethylbenzene proceeds readily over
the medium pore zeolites. Karge and his co-workers (Karge et
al., 1982; Karge et al., 1984) used ethylbenzene transalkylation
as a test reaction for the determination of acid sites in ZSM-5
and ZSM-11 and found a linear relationship between conversion and the aluminum content of the zeolite. As can be expected, ZSM-5 and ZSM-11 produced high concentrations of

p-di-ethylbenzene and no o-di-ethylbenzene (Kaeding, 1985).
This is unlike mordenite, which gave an essentially equilibrated
di-ethylbenzene composition.

Transalkylation of C_8 aromatics takes place between two
ethylbenzene molecules and between an ethylbenzene and a
xylene molecule to produce a mixture of benzene, toluene, trimethyl, methyl-ethyl, diethyl and dimethyl-ethylbenzenes
(Figure 4.12).

From the product composition, the relative rates of these
reactions over ZSM-5, mordenite and ZSM-4 have been determined (Olson and Haag, 1984). As shown in Table 4.10, the
data clearly indicate that the selectivity of ZSM-5, mordenite and
ZSM-4 is quite different for these transalkylation/disproportiona-
tion reactions. They demonstrate the pronounced effect of
zeolite pore structure on reaction selectivity. Transalkylation of
C_8 aromatics in ZSM-5 occurs predominantly between ethylbenzene molecules, rather than between ethylbenzene and
xylenes; the latter reaction occurs readily in larger pore zeolite
catalysts. The more selective ethylbenzene transalkylation in
ZSM-5 is a result of the faster diffusion rate of ethylbenzene
relative to that of m- or o-xylene.

Toluene and C_9^+ aromatics undergo transalkylation reactions
over ZSM-5 (Brennan and Morrison, 1976) to produce a mixture
of xylenes.

Table 4.11 shows the result of transalkylating a mixture of
toluene and C_9^+ aromatics. The latter, comprising predominantly
methyl and ethyl substituted alkylbenzenes, was obtained by
fractionating a catalytic reformate stream. The reaction provides
a new route to upgrade heavy aromatics to xylenes.

While xylene disproportionation is difficult with ZSM-5, the
catalyst is very effective for the disproportionation of toluene.

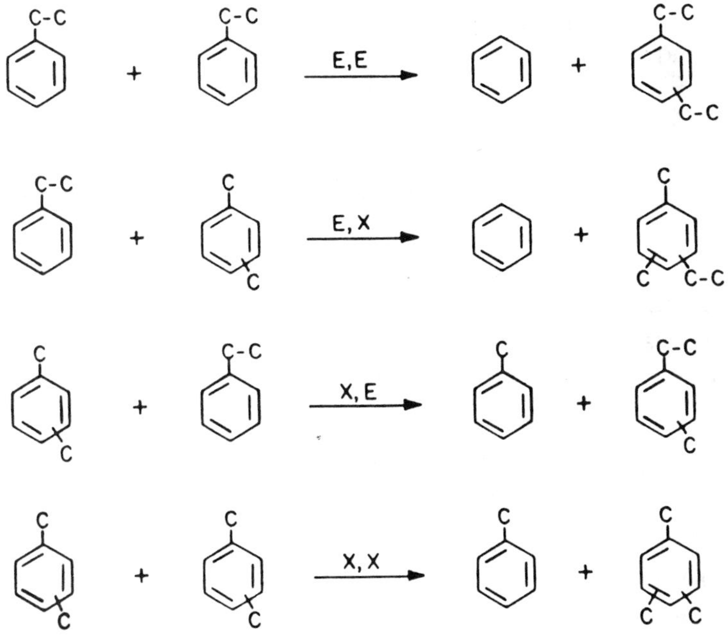

Figure 4.12 Transalkylation reactions in the ethylbenzene-xylene system. (From Olson and Haag, 1984.)

Table 4.10 Selectivity in Xylene Isomerization

Feed: 15% Ethylbenzene
      85% Xylene (63% *m*, 22% *o*)

Catalyst	% Xyl transalkylated % EB transalkylated
ZSM-4	.36
Mordenite	.19
ZSM-5	.09

Source: Olson and Haag (1984).

Table 4.11   Transalkylating a Mixture of
Toluene and C_9^+ Aromatics

Reaction Conditions:	43 atm.	
	370°C	
	4/1 molar H_2/HC	
	1 WHSV	
Wt %	Feed	Product
Nonaromatics	0.2	7.4
Benzene	—	7.2
Toluene	44.4	37.6
Ethylbenzene	0.3	—
Xylenes	4.9	26.8
C_9 Aromatics	38.9	17.4
C_{10} Aromatics	9.4	1.3
C_{11}^+ Aromatics	1.9	2.3

Source: Brennan and Morrison (1976).

Alkylation

Ethylbenzene synthesis
Zeolite catalyzed alkylation of aromatics with ethene was reported
by Wise, using rare earth exchanged faujasite (Wise, 1966), and by
Karge and his co-workers using mordenite (Becker et al., 1973).

Alkylation of benzene is accompanied by the polymerization of ethene. Hence a high molar ratio of benzene to ethene is necessary to minimize the polymerization of ethene and the formation of polyalkylated benzenes, which cause rapid catalyst deactivation (Wise, 1966).

ZSM-5 inhibits the formation of polyalkylated benzenes normally formed in significant quantities with nonshape selective catalysts. This unique property led to more stable catalysts and improves the product selectivity far superior to all previously available catalysts. The catalyst is used in the new ethylbenzene process jointly developed by Mobil and the Badger Company.

Synthesis of propylbenzenes
Alkylation of benzene with propene over HZSM-5 at below 275°C yields primarily cumene (isopropylbenzene) as expected from classical acid catalyzed alkylation.

At a 0.7 weight ratio of propene/benzene in a mixture of hexanes and heptanes, propene is completely converted, while the paraffin conversion is <5%. Most of the propene that is not consumed in alkylation goes to higher molecular weight olefins. As pressure is increased, some polyalkylated products are formed (Brennan et al., 1981) (Table 4.12).

Synthesis of long chain alkylbenzenes
The influence of structure on product distribution was also observed on other medium pore zeolites.

Young found that ZSM-12, which has a constraint index lower than that of ZSM-5, gave unique product distributions in the alkylation of benzene (Young, 1981a) with long chain olefins. They found preferential formation of 2-phenyl alkanes as shown in Table 4.13.

Table 4.12   Alkylation of Benzene with Propene 3 WHSV Benzene Blend[a], 0.5 WHSV Propene, 275°C

Pressure, atm.:	1	28	48
Propene conv., wt %	95	99	99
Benzene alkylated, wt %	10	29	43
	Products, wt %		
Methane + ethane	<0.1	<0.1	<0.1
Propane	0.4	0.3	0.3
Propene	0.7	0.1	0.1
i-Butane	0.5	0.5	0.4
n-Butane	0.2	0.3	0.5
Butenes	1.3	<0.1	<0.1
i-Pentane	0.5	0.4	0.5
n-Pentane	0.3	0.5	0.6
Pentenes	1.6	<0.1	<0.1
$C_6 + C_7$ Paraffins	69.4	68.0	65.4
C_6^+ Olefins	2.7	0.3	0.7
Benzene	17.8	14.0	11.3
Toluene	<0.1	<0.1	<0.1
C_8 Alkylbenzenes	<0.1	<0.1	<0.1
Isopropylbenzenes	3.5	3.6	10.3
n-Propylbenzenes	<0.1	<0.1	0.3
Other alkylbenzenes	1.5	7.0	9.6

[a]Benzene blend, wt %: 23% benzene, 12% n-hexane, 30% 2-methylpentane, 6% 2,3-dimethylbutane, 4% n-heptane, 20% 2,4-dimethylpentane.
Source: Brennan et al. (1981).

Table 4.13 Alkylation over ZSM-12

Benzene + dodecene, 200°C, 14.6 atm[a]

Dodecene conversion, wt %	Phenyldodecane isomer distribution, %				
	2-ϕ	3-ϕ	4-ϕ	5-ϕ	6-ϕ
54	92	8	0	0	0

Phenol + 1-octanol, 250°C, 16.6 atm[b]

Octylphenol isomer distribution

Octanol conversion wt %	Para		Ortho		Others	Ratio, octylphenol to dioctylphenol
	2-Octyl	3-Octyl	2-Octyl	3-Octyl		
100	37	41	16	0	6	135

[a]*Source:* Young (1981a).
[b]*Source:* Young (1981b).

$$R-CH = CH-CH_3 \;\rightleftharpoons\; R = CH_2 \;\rightleftharpoons\; R-CH = CH-R$$

$$\downarrow$$

$$CH_3-CH_2-R$$

Cymene synthesis

Alkylation of toluene with propene over ZSM-12 was studied by Kaeding and his co-workers (1980). At 230-240°C and 33 atm, feeding a 6.25 molar ratio of toluene to propene at 6.1 WHSV, 90% to 95% of the propene fed was converted. The selectivity to isopropyltoluene (cymene) exceeded 95%. Table 4.14 shows the isomer distribution.

Diarylalkanes

The formation of diarylalkanes and polyalkylated benzenes during the alkylation of benzene or toluene over the medium pore zeolites is sterically hindered.

However, under more severe reaction conditions, these sterically hindered reactions can take place to some extent to yield unique products only possible with shape selective zeolites. For example, Sato et al. (1986) showed that in the alkylation of toluene with ethene over HZSM-5 at 450°C and 1 atm, 2.1% of the alkylated product boiled above 250°C. The fraction boiled

Table 4.14   Isomer Distribution of Isopropyltoluenes

Reaction	ortho	meta	para
Temperature, 230°C	5	65	30

Source: Kaeding et al. (1980).

between 275 and 320°C comprised 85% of diarylalkanes with the composition shown in Table 4.15. These compounds have been shown to be excellent electrical insulating oils.

Alkylbenzenes from paraffins/aromatics mixtures

Paraffins are normally not suitable as alkylating agents for aromatics, although in some instances low yields of alkylaromatics have been obtained with $AlCl_3$ as an acid catalyst. With medium pore zeolites, such as ZSM-5, this reaction occurs with surprising ease and good yield.

When paraffins are cracked over ZSM-5 in the presence of benzene and toluene, the aromatic molecules rapidly "scavenge" the cracked fragments (Garwood and Chen, 1980).

Table 4.15 High-Molecular-Weight Products
from Alkylating Toluene with Ethene

Catalyst	ZSM-5, $SiO_2/Al_2O_3 = 60$
Temperature	450°C
Pressure	1 atm
WHSV	4.5
Ethene/toluene, molar	0.2

Composition of 275–320°C product

	Wt %
Diarylalkanes	85.0
C_nH_{2n-14}	
$n = 14$	15.3
$n = 15$	43.8
$n = 16$	25.9
Others	15.0
Total	100.0

Source: Sato et al. (1986).

This is demonstrated by charging a mixture of *n*-octane and benzene over HZSM-5 at 49 atm and 315°C (Table 4.16). A carbon
balance shows the fate of the octane converted (Table 4.17).
Note that negligible amounts of new rings were made under
these conditions. This is in contrast to the case where *n*-octane
alone was converted under similar conditions (Table 4.8); 17%
of the converted octane appeared as aromatics formed by a series
of reactions described earlier as M2-Forming reactions.

     With benzene in the feed, the most predominant products
are C_3–C_5 paraffins and propyl- and butyl-benzenes. Their formation is readily explained by an initial cracking of *n*-octane to

Table 4.16   1/1 Weight Blend n-Octane/Benzene
over ZSM-5 315°C, 49 atm, 4 LHSV

	Products, wt %
Methane + ethane	<0.1
Propane	7.3
Propene	<0.1
Butanes	10.7
Pentanes	5.7
C_6 + C_7 Paraffins	2.4
Octane	6.4
Benzene	24.7
Toluene	0.5
C_8 Alkylbenzenes	1.3
Isopropylbenzene	8.5
n-Propylbenzene	11.2
Other C_9 alkylbenzenes	0.5
n + sec-Butylbenzene	7.1
$tert$-Butylbenzene	1.9
Other C_{10} alkylbenzenes	1.7
C_{11} + alkylbenzenes	10.1
	100.0

Source: Garwood and Chen (1980).

Table 4.17   Fate of Converted
n-Octane

	wt %
Light paraffins	65
Alkylbenzene side chains	34
Aromatic ring carbon	<1

Source: Garwood and Chen (1980).

light C_3^+ paraffins and olefins, followed by alkylation of benzene by the propene and butenes. Alkylation by pentene is sterically hindered. The primary alkylation products are those expected from classical acid-catalyzed alkylation: isopropylbenzene and *sec*-butylbenzene. Under more severe reaction conditions, they first undergo secondary isomerization reactions to form *n*-propylbenzene and *n*-butylbenzene. This is followed by dealkylation, olefin isomerization/transmutation, and realkylation reactions in addition to disproportionation reactions to produce the spectrum of aromatic products.

Cracking longer chain paraffins over ZSM-5 produces primarily C_3 and C_4 olefins which are scavenged by benzene. Thus, with *n*-hexadecane the weight percent of converted paraffin that ends up as alkyl aromatics reaches about 75% (Figure 4.13).

When both benzene and toluene are present in the reaction mixture, they are alkylated by the cracked fragments. Toluene is intrinsically more reactive for alkylation than benzene. According to the selectivity relationship developed for non-sterospecific catalysts by Stock and Brown (1959), the relative rate constants, k(Toluene)/k(benzene), for alkylation should be about 1.4. This relationship was found to hold for both homogeneous and heterogeneous catalyst systems (Brown and Smoot, 1956; Venuto, 1967). However, over ZSM-5, as shown by the data obtained at 35 atm, (Table 4.18), the observed ratio of k(toluene)/k(benzene) is about half of that for the nonstereospecific case,

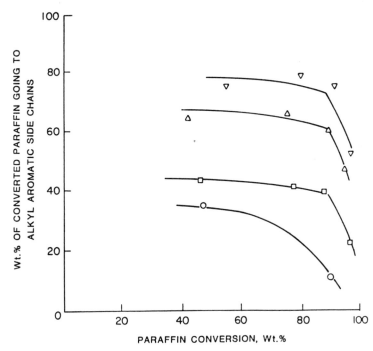

Figure 4.13 Conversion of *n*-paraffins to alkyl aromatic side chains, 48 atm, 315°C, severity controlled by space velocity. (From Garwood and Chen, 1980.)

Table 4.18   Rate of Aromatics Alkylation[a]

	Temperature, °C	
Conversion (wt %)	315	340
Benzene	25	41
Toluene	18	31
k(toluene)/k(benzene)	0.69	0.70

[a]35 atm. No hydrogen was used.
Source: Chen and Garwood (1978b).

suggesting that the channel openings in ZSM-5 either impose a stereospecific effect on the alkylation reactions or that the rate of alkylation is inhibited by the slow diffusion of some of the dialkylbenzene isomers (Chen and Garwood, 1978b).

This interpretation is consistent with the observation that as the reaction pressure is reduced, the observed ratio of k(toluene)/k(benzene) increases toward that of the intrinsic ratio (Table 4.19).

Dealkylation of Alkylaromatics

The dealkylation of alkylaromatics over ZSM-5 as in the case of alkylation is accompanied by secondary reactions, that is, olefin isomerization/transmutation and realkylation reactions in addition to disproportionation and hydrogen transfer reactions to produce a spectrum of aromatics and gaseous products. Table 4.20 shows the product distribution when n-propylbenzene is dealkylated at 28 atm in 3/1 mol ratio of hydrogen to hydrocarbon.

Sato et al. (1985) showed that when the dealkylation reaction was carried out in an inert atmosphere with MgO modified LiZSM-5, secondary reactions of the dealkylated olefin could be avoided. Table 4.21 compares the composition of gaseous product of HZSM-5 with that of the lithium exchanged ZSM-5 when cymene was dealkylated at 1 atm and 250°C.

Table 4.19   Effect of Pressure on the Relative
Rate of Aromatics Alkylation[a]

Conversion, wt %	Pressure, atm	
	1	48
Benzene	24	65
Toluene	30	52
k(toluene)/k(benzene)	1.30	0.74

[a]~280°C, propene/(benzene + toluene) = 1/1 molar.
Source: Garwood and Chen (1980).

Table 4.20   Dealkylation of n-Propylbenzene
over HZSM-5[a]

	Temperature, °C	
	315	370
$C_1 + C_2$	0.2 (wt %)	0.4 (wt %)
C_3	4	7
C_4	6	8.5
C_5^+	4	5
Benzene	38	45
Toluene	2	4
C_8 Aromatics	4	9
C_9 Aromatics	24	9
C_{10}^+ Aromatics	18	12

[a]28 atm, 3/1 molar ratio of hydrogen to hydrocarbon.
Source: Brennan and Morrison (1976).

Table 4.21   Dealkylation of Cymene, 1 atm

Catalyst	Composition of gaseous product, mol %					
	Ethene	Propene	Propane	$C_4^=$	$C_5^=$	$C_6^=$
MgO modified LiZSM-5, at 500°C	0	95	0	2	1	2
HZSM-5, at 250°C	1	11	2	46	28	12

Source: Sato et al. (1985).

Table 4.22   Hydrodealkylation of Alkylbenzenes over Pt/ZSM-5 Catalyst, 1 atm, 400°C

	Feed	Product, wt %
Methane		0.01
Ethane		24.69
Ethene		0.02
Propane		0.59
Butanes		0.04
Benzene		23.62
Toluene		8.72
Ethylbenzene		3.09
Xylenes		16.55
Cumene	0.9	—
Ethyltoluene	9.0	0.45
Trimethylbenzenes	10.0	9.52
Diethylbenzenes	47.0	0.52
Ethylxylenes	33.1	12.18
Total	100.0	100.00

Source: Onodera et al. (1982).

By adding a noble metal function to the catalyst, alkylbenzenes with C_2^+ alkyl groups can be selectively hydrodealkylated
to a mixture of benzene and methyl-substituted benzenes (Bonacci
and Billings, 1976; Onodera et al., 1982) with ethane as the major
by-product. For example, a feed containing 9% ethyltoluene, 10%
trimethylbenzene, 47% diethylbenzene, and 33.1% ethylxylene
was dealkylated at 1 atm and 400°C over a Pt/ZSM-5 catalyst to
yield a mixture of BTX and ethane, as shown in Table 4.22.

Paring Reactions

The paring reaction, forming light isoparaffins, particularly isobutane, from polymethylbenzenes, is described by Sullivan et al. (1961),
to occur by a mechanism involving ring contraction and expansion
and isomerization.

To explain the observed "scrambling" of side chains around the aromatic ring to form a wide spectrum of polymethyl and ethyl benzenes from the initial alkylation products, isopropyl benzene and *sec*-butylbenzene, a "reverse" paring reaction was proposed (Chen et al., 1979a) as an alternative to dealkylation/olefin transmutation/realkylation pathway. However, the relative importance of these reaction schemes remains to be investigated.

Para-selective Reactions

Molecules such as p-dialkylbenzenes diffuse much more rapidly into medium pore zeolites than their ortho- and meta-isomers do. For example, the diffusion rate of p-xylene in ZSM-5 is 10^5 times greater than 1,3,5-trimethylbenzene (mesitylene) (Olson et al., 1981). Through adjustment of the acid activity and the diffusion parameters (such as crystal size and/or chemical treatment) of the zeolite, such that the reactions involving the m- and o-isomers are severely diffusion-constrained, high para-selectivity has been achieved in the alkylation of toluene with methanol (Chen, 1977, 1978; Chen et al., 1979b; Haag and Olson, 1978; Kaeding et al., 1981a). Similarly, selective formation of p-ethyltoluene was made possible by alkylating toluene with ethene (Kaeding and Young, 1978; Kaeding et al., 1983), p-di-ethylbenzene by alkylating ethylbenzene with ethene (Ishida and Nakajima, 1986) and the preferential formation of p-cymene (1-methyl-4-isopropylbenzene) was made possible by the alkylation of toluene with propene. A ratio of greater than 10 of cymene to n-propyltoluene has been achieved with ZSM-5 (Dwyer and Klocke, 1977).

In addition to alkylation reactions, selective toluene disproportionation (STDP) has been achieved to yield p-xylene as the predominant xylene isomer (Table 4.23) (Chen et al., 1979b; Kaeding et al., 1981b). Selective dealkylation of p-dialkylbenzenes in a dialkylbenzene mixture has also been achieved (Table 4.24) (Sato et al., 1985).

Haag and his co-workers (Olson and Haag, 1984) developed a correlation between observed para-selectivity in toluene

Table 4.23 Toluene Disproportionation over Modified ZSM-5 Catalysts

Modifier:	Phosphorous		Boron			Magnesium
Temperature, °C	555	600	600	650	700	550
Pressure, atm	1	1	1	1	1	1
Toluene, WHSV	2.1	2.1	2.8	2.8	2.8	3.5
conversion, %	0.3	0.8	5.4	9.7	13.1	10.9
Time on stream, hr	0–1	1–2	1.0	1.0	1.0	1.0
Product, wt %						
benzene	0.14	0.34	2.9	5.4	8.4	4.9
toluene	99.66	99.18	94.6	90.4	86.85	89.2
p-xylene	0.10	0.19	1.85	3.1	3.9	5.2
m-xylene	0.05	0.14	0.5	0.8	0.45	0.6
o-xylene	—	0.05	0.15	0.2	0.1	0.1
C_9^+ aromatics	—	—	0	0.1	0.3	0
Total	100.00	100.00	100.00	100.0	100.00	100.0
Xylene isomers, %						
para	67	50	74	74	87	88
meta	33	37	21	20	10	10
ortho	—	13	6	6	3	2

Source: Kaeding et al. (1981b).

Table 4.24   Selective Dealkylation of para-Cymene over MgO
Modified LiZSM-5

Feed: 32.9% p-cymene, 63.6% m-cymene, 3.5% o-cymene
Pressure: 1 atm, temperature: 500°C

Contact time, sec	Conversion, wt %		Propene purity, mol %
	para	meta and ortho	
4.0	68.6	1.5	97
10.2	91.2	4.3	93

Source: Sato et al. (1985).

disproportionation and a model based on the interplay between
diffusion and reaction. In this model, shown in Figure 4.14,
toluene diffuses into the zeolite with a diffusivity, D. It undergoes disproportionation to benzene and a mixture of xylenes
(P, M, O), which equilibrate rapidly in ZSM-5. At short residence
times, the primary products (P, M, O) leaving the zeolite have an
isomer distribution determined by the relative rate of isomerization, the diffusion coefficient of the respective isomers $(D_p, D_m$
and $D_o)$, and the radius of the zeolite crystal. A more rigorous
mathematical analysis based on similar premises was reported
recently by Wei (1982), who showed that regardless of the
intrinsic product selectivity, the observed isomer distribution is
altered as a result of unequal diffusion rates and isomerization
rate of the isomers. The enrichment of p-xylene is made possible by the much greater diffusion coefficient of p-xylene. It
can be shown that when the rate of isomerization is much faster
than that of diffusion, that is, when the effectiveness factor,
square root of kR^2/D, is greater than 10, the observed isomer
distribution at very low conversions will contain better than 90%
p-xylene regardless of the intrinsic isomer distribution.

    At longer residence times, the primary product on re-enter-
ing the zeolite would reisomerize to form secondary products.

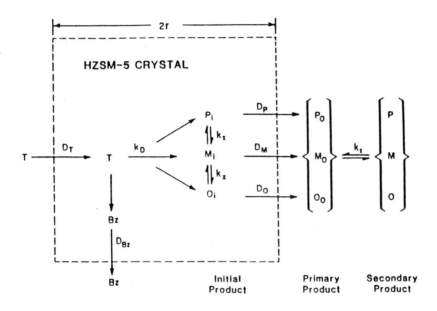

Symbols:

D = Diffusion Coefficient
k = Rate Constant
r = Crystal Radius
Bz = Benzene
T = Toluene
P,M,O = Para-, Meta, and Ortho-xylene

Subscripts:

I = Isomerization
D = Disproportionation
i = Initial Product
o = Primary Product

Figure 4.14   Model for STDP.   (From Olson and Haag, 1984.)

The composition of the secondary product will thus be a function of conversion. Theoretical analysis (Wei, 1982) shows that the relation between para-selectivity and conversion depends on the competitive rates of alkylation (disproportionation) and isomerication, or the ratio of $k_a/k_i^{1/2}$. Figure 4.15 shows that, when the experimental data are fitted to the theoretical curves, $k_a/k_i^{1/2}$ values of 1 and 10 are obtained for toluene disproportionation and toluene alkylation respectively.

In addition to the above examples, similar kinds of para-selective reactions involving nonhydrocarbons can be found in the patent literature, to name a few, the synthesis of p-alkylanisoles (Young, 1983) and p-alkylphenols (Toray Ind., 1986) by alkylation of methoxybenzene and phenol respectively; and p-di-halo-benzenes (Asahi Chem., 1984; Ihara Chem., 1985b) and p-halo-alkoxybenzenes (Ihara Chem., 1985a) by halogenation of benzene and alkoxybenzenes, respectively. Many more can be expected.

D. Naphthene Reactions

Naphthenes are less reactive than paraffins and olefins.

For example, the conversion of methylcyclopentane at 345°C is only 23% compared to over 80% conversion for n-octane. However, once the ring is opened, the resultant olefins undergo a series of M2-Forming reactions producing the wide spectrum of aromatic and paraffinic products (Table 4.25). Because of their molecular size, bicyclic naphthenes, such as decalin, are much less reactive over ZSM-5.

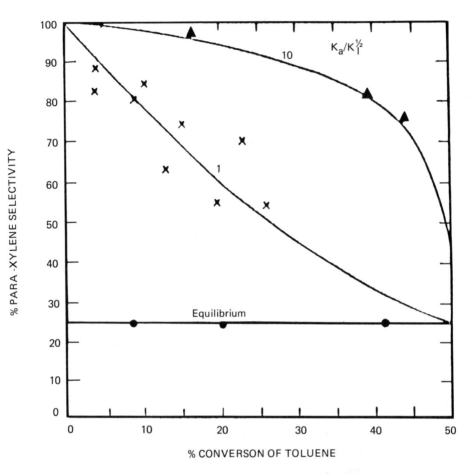

Figure 4.15   The relation between the percentage para-selectivity
and percentage toluene conversion in toluene disproportionation.
Theoretical curves for $K_a/K_I^{1/2}$ of 1 and 10: ($\bullet$) HZSM-5 disproportionation; (X) MgZSM-5, disproportionation; ($\blacktriangle$) PZSM-5,
methanol alkylation. (From Wei, 1982.)

Table 4.25   Products from Methylcyclopentane Cracking
35 atm, 345°C, 1 LHSV

	Wt %
Methane	<0.1
Ethane	0.3
Propane	3.6
Propene	0.1
i-Butane	1.3
n-Butane	1.3
Butenes	<0.1
i-Pentane	0.2
n-Pentane	0.3
C_6 Paraffins	5.0
C_7 Paraffins	0.5
C_8 Paraffins	0.3
Benzene	<0.1
Toluene	1.7
C_8 Alkylbenzenes	3.1
C_9 Alkylbenzenes	2.7
C_{10} Alkylbenzenes	0.7
C_{11} Alkylbenzenes	<0.1
Methylcyclopentane	77.5
Dimethylcyclopentane	0.4
Cyclohexane	0.2
Methylcyclohexane	0.3

Source: Chen et al. (1979a).

E. Reactions of Oxygen-Containing Compounds

Alcohols and Ethers

Hydrocarbons from methanol and dimethyl ether

The chemistry of methanol/dimethylether to hydrocarbons reactions over medium pore zeolites has been comprehensively reviewed by Chang (1983) recently. There is still disagreement over the mechanism of initial C-C bond formation and whether the primary hydrocarbon from methanol/dimethyl ether is ethene or propene or both (Espinoza and Mandersloot, 1984; Chu and Chang, 1984; Wu and Kaeding, 1984). While positive identification of the primary product remains elusive, recent data (van den Berg et al., 1980; Haag et al., 1982b) lend support to the belief that ethene is the intrinsic primary hydrocarbon product. But if the rate of the secondary reaction, i.e., the autocatalytic methylation of ethene-to-propene reaction (Chen and Reagan, 1979a; Ono and Mori, 1981) is so much faster than the rate of desorption or diffusion of the primary product, then under diffusion/desorption disguise, propene may appear to be the primary product.

Subsequent reactions include the redistribution of olefins via oligomerization/transmutation reactions to a thermodynamic equilibrium composition, including ethene. Radiotracer studies proved that the ethene found in the final product is not the original primary product (Dessau and LaPierre, 1982). At low conversions, the olefin composition also correlates well with a stepwise chain-growth mechanism (Wu and Kaeding, 1984).

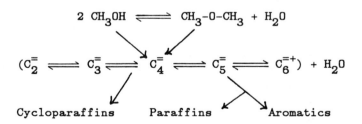

Other than the effect of water and oxygenates, the chemistry of the conversion of olefins to other hydrocarbons is essentially analogous to that described earlier. Figure 4.16 shows the reaction path for methanol conversion to hydrocarbons over HZSM-5 at 371°C, and the shaded area represents a region where the unconverted oxygenates and aromatics coexist leading to ring alkylation and the formation of polyalkylbenzenes. As expected, the nature of the aromatics formed from methanol is dictated by the structural effects of the zeolite.

This overlapping region may be expanded by increasing the reaction pressure and contracted by lowering the partial pressure of methanol in the feed. At higher pressures, preferential formation of durene (1,2,4,5-tetramethylbenzene) far exceeding its equilibrium value has been observed. Thus, a new process for the manufacture of durene is possible with medium pore zeolites (Chang et al., 1975b; Fowles and Yan, 1985).

Preferential production of light olefins from methanol may be achieved in several ways, including decreasing residence time, decreasing partial pressure of the oxygenates in the feed, and operating at less than 100% conversion of the oxygenates (Chang et al., 1979).

Alternatively, reducing the acidity of the zeolite and operating at much higher temperatures (Chang et al., 1984) uncouples the olefin formation reactions from the aromatization reactions. C_2 to C_5 olefin yields as high as 80% have been obtained (Figure 4.17). This approach has led to the development of Mobil's Methanol to Olefin Process (MTO) (Gould et al., 1986).

Ethene from methanol/dimethyl ether
The selective conversion of methanol/dimethylether to ethene represents an interesting challenge to shape selective catalysis. Ethene yields over the medium pore zeolites are seldom over 10% of the total hydrocarbon product even when the reaction severity is chosen to maximize light olefins (Chang et al., 1984).

Diluting methanol with water shifts the methanol/dimethylether equilibrium toward methanol and is reported to have a dramatic effect on increasing ethene selectivity (Caesar and Morrison, 1978). Ethene selectivity as high as 28% has been

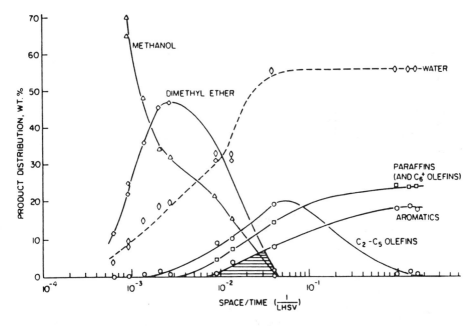

Figure 4.16   Reaction path for methanol conversion to hydrocarbons over HZSM-5 (371°C). (From Chang and Silvestri, 1977.)

observed at a 10/1 molar ratio of water to methanol and 50%
conversion (Figure 4.18).

A number of investigators (Chang et al., 1977; Singh and
Anthony, 1979; Singh et al., 1980; Ceckiewicz, 1981; Wunder
and Leupold, 1980, and Inui et al., 1981) have used small pore
zeolites, such as chabazite/erionite, zeolite T, TMA-erionite/
offretite, ZK-5, and ZSM-34 for methanol conversion. Represen-
tative data reported by Chang et al. are shown in Table 4.26.
Although they generally produce more ethene when compared to
medium pore zeolites, the reaction is always accompanied by
rapid coke deactivation (Givens, et al., 1978). This would be
expected from structural considerations of these zeolites
(Derouane, 1985).

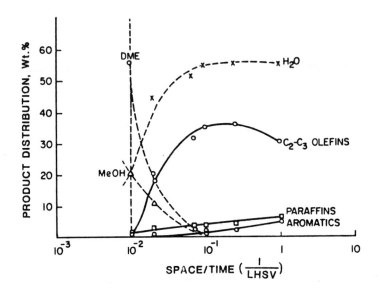

Figure 4.17   MeOH conversion over HZSM-5 (SiO_2/Al_2O_3 = 500) 500°C, atmospheric pressure. (From Chang et al., 1984.)

However, more recently, Kaiser (1985) reported that with phosphate-based molecular sieves, analogs of small pore zeolites, such as erionite (SAPO-17 and ALPO4-17) and chabazite (SAPO-34 and ALPO4-17), complete conversion of methanol was maintained for more than 6 hours. Unfortunately, operating at 100% conversion provides no information on the rate of deactivation. Further study is necessary to determine whether any difference in deactivation between these two types of molecular sieves can be attributed to either the difference in the density or the strength of their acid sites.

In any case, these studies on small pore catalysts provided a clue to selective ethene production, that is, compared to medium pore zeolites, small pore zeolites gave a higher ratio of ethene to C_3^+ olefins, probably the result of steric inhibition on the methanol alkylation reaction. Therefore the pore size of the zeolite and/or diffusivity must play a role in giving high ethene selectivity.

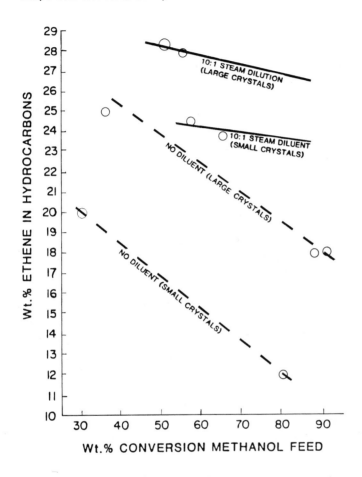

Figure 4.18   Effect of water dilution on methanol conversion to olefins. (From Caesar and Morrison, 1978.)

Table 4.26   Methanol Conversion over Small Pore Zeolites

	Erionite[a]	Zeolite T	Chabazite	ZK-5
Reaction conditions:				
temperature, °C	370	341–378	538	538
pressure, atm	1	1	1	1
LHSV (WHSV), h^{-1}	1	(3.8)	[b]	[b]
conversion, %	9.6	11.1	100	100
Hydrocarbons, wt %:				
CH_4	5.5	3.6	3.3	3.2
CH_2H_6	0.4	0.7	4.4	0
C_2H_4	36.3	45.7	25.4	21.4
C_3H_6	1.8	0	33.3	31.8
C_3H_6	39.1	30.0	21.2	13.5
C_4H_{10}	5.7	6.5 ⎫	10.4	22.6
C_4H_6	9.0	10.0 ⎭		
$C_5{}^+$	2.2	3.1	2.0	7.5

[a]De-aluminized; $SiO_2/Al_2O_3 = 16$.
[b]Pulse microreactor, 1 μL MeOH in He, 500 h^{-1} GHSV.
Source: Chang (1983).

Olson and Haag (1984) showed that it is possible to reduce the diffusivity by reducing the pore volume. In a series of experiments, Olson et al. reduced zeolite sorption capacity by impregnation with magnesium oxides. Some success has been reported on increasing ethene selectivity by the addition of other inorganic oxides to the medium pore zeolites. For example, Kaeding and Butter (1980) reported improved ethene selectivity when the ZSM-5 was impregnated with phosphorus compounds.

Similar effects have been observed following impregnation of ZSM-5 with organic silicone compounds (Figure 4.19). When an organic silicon-containing compound, such as dimethylsilane, was adsorbed into the pores of the zeolite, and after calcining, a reduction of zeolite sorption capacity was noted.

ZSM-5 impregnated with antimony oxide (SbO) represents another interesting approach. The zeolite can sorb more than 20 wt % of Sb_2O_3. The impregnated zeolite showed high initial ethene selectivity (Table 4.27).

However, as the antimony oxide is reduced during the reaction, it leaves the zeolite and the zeolite reverts to its unmodified form giving low ethene yield. High ethene selectivity can be maintained by cofeeding an oxidant, such as hydrogen peroxide with the methanol (Chen and Reagan, 1977).

It is interesting to note that various approaches all lead to the same upper limit of about 33% ethene at 60% conversion of methanol in a 20 wt % methanol (7:1 molar ratio of water: methanol) solution. Only by the addition of a dehydrogenation

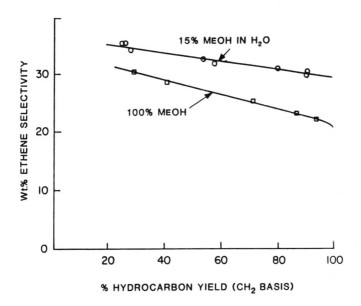

Figure 4.19 Conversion of methanol to ethene over silica-modified HZSM-5, 340–360°C. (From Rodewald, 1979.)

Table 4.27   Methanol Conversion over
Antimony Modified ZSM-5 Catalysts

Temperature, °C	300	350
Aliphatics, wt %		
H_2	.05	.34
CO	.86	1.70
CO_2	.20	.75
Methane	.58	1.07
Ethane	.23	.44
Ethene	35.90	27.05
Propane	4.55	3.32
Propene	36.92	29.79
Butanes	4.24	2.11
Butenes	13.55	10.18
C_5's	2.92	3.21
C_6's	0	1.14
C_7^+'s	0	4.24
Aromatics, wt %		
Benzene	0	.37
Toluene	0	1.77
Xylenes	0	8.61
C_9 Alkylbenzenes	0	2.50
C_{10} Alkylbenzenes	0	1.41

Source: Butter (1976).

function coupled with inorganic oxide modifiers, for example, ZnPd with MgO, could the ethene selectivity be further increased to as high as 47% under the same operating conditions (Figure 4.20).

Hydrocarbons from ethanol and higher alcohols
The conversion of ethanol and higher alcohols to hydrocarbons is less difficult to understand than methanol, because the dehydration of C_2^+ alcohols to light olefins is well known.

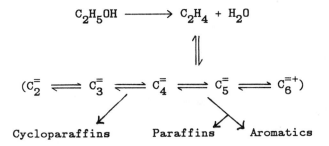

Figure 4.21 shows the reaction pathway for ethanol conversion to hydrocarbons over HZSM-5 at 371°C (Chen, 1982; Chang, 1985).

The thermochemistry of ethanol reactions is different from methanol conversion, which is highly exothermic. Although the overall heat of reaction for ethanol is less exothermic than the methanol reactions, the initial step of dehydration of higher alcohols is highly endothermic, followed by the highly exothermic olefinic reactions. This makes temperature control a problem in commercial reactors other than a fluidized bed (Chen, 1983).

Alkylation of aromatics is again an important factor in the overall reaction scheme, as evidenced by the preponderance of ethyl-substituted side chains in the liquid product (Table 4.28).

Methyl tert-butyl ether (MTBE) synthesis
Methyl *tert*-butyl ether, a high octane (115–135 RON) gasoline blending component (Pecci and Floris, 1977) can be synthesized

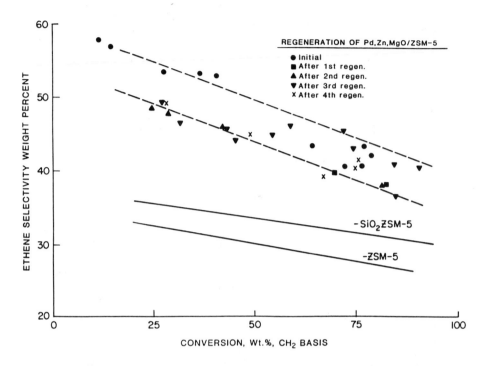

Figure 4.20   Conversion of methanol to ethene over modified ZSM-5 catalysts. (From Chen and Reagan, 1979b, 1981.)

from methanol and isobutene over the medium pore zeolite. Compared to Amberlyst 15, a microreticular hydronium ion-exchanged resin currently in commercial use, HZSM-5 and HZSM-11 (Chu and Kuehl, 1987) can tolerate higher temperatures, although the reaction is equilibrium-limited. They also give 100% selectivity for MTBE without the need of having an excess methanol. The latter is attributed to the selective enrichment of methanol concentration in the zeolite pores.

Carbonylation of methanol
Nagy (1979) in his ^{13}C-nmr investigation of the conversion of methanol over ZSM-5 in the presence of 130 toor of CO found

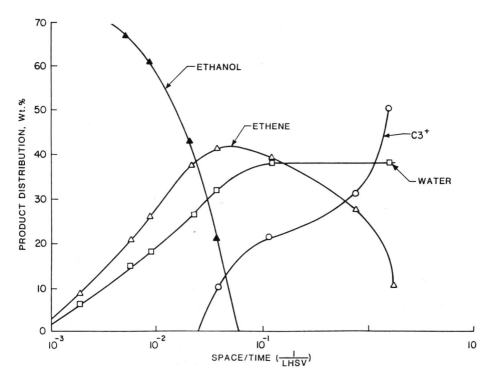

Figure 4.21   Reaction path for ethanol conversion to hydrocarbon over HZSM-5 at 371°C. (From Chen, 1982.)

that CO had an negligible effect on the rate of conversion of methanol although some carbon monoxide appeared to be incorporated into the products. The carbonylation activity of ZSM-5 was later confirmed by Feitler (1986). In the latter study, the reaction was carried out at 69 atm and 355°C, with CO and methanol being fed at a molar ratio of 33:1. Methanol was converted to hydrocarbons and oxygenates with the product distribution shown in Table 4.29.

Table 4.28   Composition of Hydrocarbon Components for
Methanol and Ethanol Feedstocks

Feedstock	Methanol	Ethanol
Methane	1.14	0.28
Ethane	0.50	1.78
Ethene	0.03	0.01
Propane	5.70	12.17
Propene	0.22	0.15
n-Butane	3.24	7.75
i-Butane	8.69	10.25
Butenes	1.08	0.68
n-Pentane	1.63	3.77
i-Pentane	11.13	9.39
Pentenes	2.09	0.77
Cyclopentane	0.33	0.35
Methylcyclopentane	1.57	1.09
n-Hexane	0.83	1.15
Methylhexanes	4.61	1.77
i-Hexanes	11.72	5.62
Hexenes	1.49	0.38
n-Heptane	0.21	0.22
i-Heptanes	0.50	0.45
C_7-Olefins	1.91	0.45
Dimethyl-N5	1.82	1.14
n-Octane	0.04	0.00
i-C_8-P + O + N_5 + N_6	6.73	3.01
n-Nonane	0.14	0.07
i-C_9-P + O + N_5 + N_6	2.43	0.86
n-Decane	0.00	0.00
i-C_{10}-P + O + N_5 + N_6	0.77	0.17
Unknown (hydrocarbon aromatics)	0.11	0.30

Table 4.28   (Continued)

Feedstock	Methanol	Ethanol
Benzene	0.23	0.93
Toluene	2.22	6.95
Ethylbenzene	0.64	2.00
($p + m$)-Xylenes	6.95	7.44
o-Xylene	1.90	2.12
Trimethylbenzenes	6.86	3.70
Methyl-ethyl-benzenes	2.74	4.37
Other C_3 Alkylbenzenes	3.71	0.16
1,2,4,5-Tetramethylbenzene	3.71	0.16
1,2,3,5-Tetramethylbenzene	0.24	0.11
1,2,3,4-Tetramethylbenzene	0.11	0.29
Other C_{10} alkylbenzenes	1.65	3.27
C_{11}-Alkylbenzenes	0.54	1.53
Naphthalenes	0.07	0.35
Unknowns (all other)	1.33	2.52
(Total C_4 + gasoline)	92.41	85.61

Source: Chen (1983).

Aldekydes and Ketones

Double bond isomerization
Hoelderich (1986) reported recently that the double bond shift
of olefinic compounds containing functional groups, such as
2-ethylacrolein to *trans*-2-methyl-butenal (tiglaldehyde) can be
achieved over low acidity medium pore zeolites with excellent
selectivity.

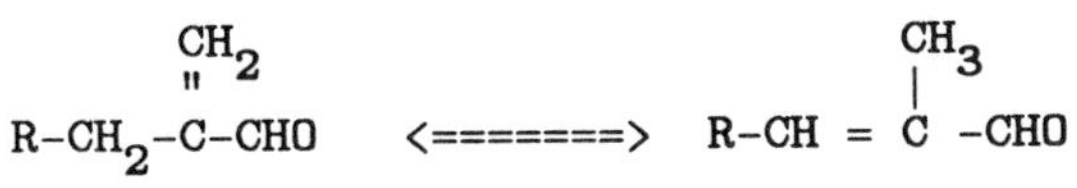

$$R-CH_2-\underset{\underset{CH_2}{\|}}{C}-CHO \quad <======> \quad R-CH = \underset{\underset{CH_3}{|}}{C} -CHO$$

Table 4.29   Evidence of Carbonylation Activity of ZSM-5

Reaction conditions: 69 atm, 355°C			
HZSM-5, SiO_2/Al_2O_3	20	52	52	46
WHSV (MeOH)	2.4	0.24	0.12	2.4
MeOH conv.	6.3	23	100	33
CO/MeOH	33	33	33	—
N_2/MeOH	—	—	—	33
Selectivity,[a] mol %			
Acetic acid + methylacetate	23	10	5	0
Ethene	24	15	10	38
Propane	5	7	7	7
Propene	7	6	5	16
C_4^+	22	53	70	37

[a]Selectivity defined as

$$\frac{\text{Moles product} \times \text{number of MeOH-based carbon in molecules}}{\text{Total moles product carbon from MeOH}} \times 100$$

Source: Feitler (1986).

Skeletal isomerization
Medium pore zeolites catalyze the interconversion of aldehydes and ketones by skeletal isomerization (Hoelderich et al., 1985). For example, isobutyladehyde is isomerized to methylethylketone at 300–500°C over ZSM-5, with better than 60% selectivity (Linstid and Koermer, 1985).

Condensation

The acid catalyzed condensation of acetone to 1,3,5-trimethyl-benzene (mesitylene) is a well-known reaction. Chang and Silvestri (1977) showed that the molecular size constraint imposed by the medium pore zeolites has a decisive effect on the product formation of this reaction. Thus, instead of the bulkier mesitylene being made, acetone is selectively converted to iso-butene over ZSM-5 under mild reaction conditions (Chang et al., 1981).

Huang and Haag (1982) showed that with the addition of a metal, for example, palladium, to the medium pore zeolites, the reaction intermediate, mesityl oxide, can be hydrogenated to form methylisobutyl ketone (MIBK) in high yields.

$$2(CH_3)_2CO \xrightarrow{-H_2O} (CH_3)_2C=CHCOCH_3 \xrightarrow{H_2} (CH_3)_2CHCH_2COCH_3$$

Acetone   Mesityl oxide   MIBK

Hagen (1984) reported that β-unsaturated aldehydes can be readily prepared from formaldehyde and aldehydes of the formula RCH_2CHO (where R = H, alkyl, aryl, aralkyl, cyclo-alkyl, or alkylaryl) via aldol condensation over medium pore zeolites.

$$R-CH_2-CHO + H_2CHO \longrightarrow CH_2 = CR - CHO$$

The acid-catalyzed condensation of formaldehyde and iso-butene, known as the Prins reaction, was studied by Chang et al. (1981) over the medium pore zeolites. Compared to nonshape selective catalysts, the molecular size constraint imposed by the medium pore zeolites was found again to inhibit the formation of bulkier cyclic condensation products, such as 4,4-dimethyl-1,3-dioxane. The selective production of 2-methylbutenol provides a simpler route to isoprene.

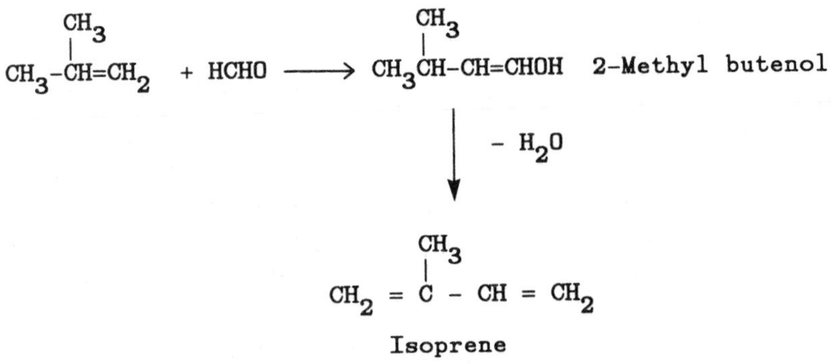

$$
\underset{\text{}}{CH_3-\overset{\overset{\displaystyle CH_3}{|}}{CH}=CH_2} \;+\; HCHO \;\longrightarrow\; \underset{\text{}}{CH_3\overset{\overset{\displaystyle CH_3}{|}}{CH}-CH=CHOH} \quad \text{2-Methyl butenol}
$$

$$
\Big\downarrow \;-\;H_2O
$$

$$
CH_2 = \overset{\overset{\displaystyle CH_3}{|}}{C} - CH = CH_2
$$

Isoprene

Dehydration

More recently, Hoelderich et al. (1985b) reported the use of low acidity ZSM-5 to dehydrate aldehydes such as 2-methylbutanal and isovaleraldehyde to isoprene.

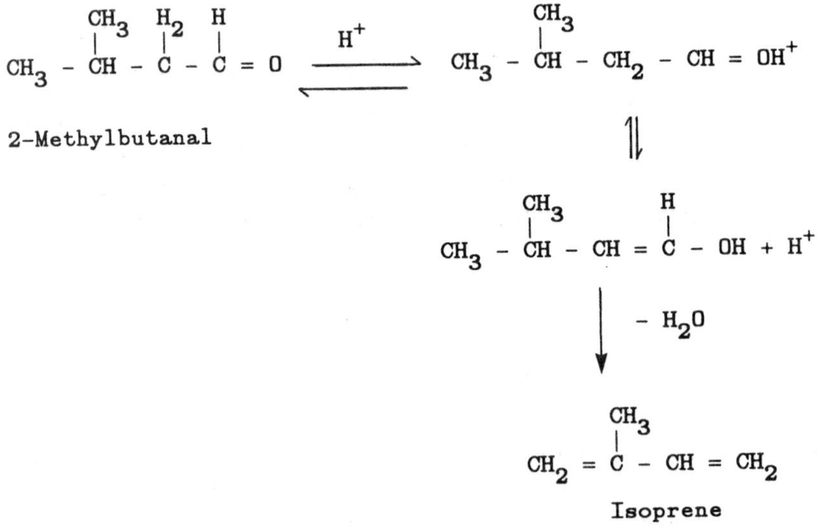

$$
CH_3 - \overset{\overset{\displaystyle CH_3}{|}}{CH} - \overset{\overset{\displaystyle H_2}{|}}{C} - \overset{\overset{\displaystyle H}{|}}{C} = O \; \underset{\longleftarrow}{\overset{H^+}{\longrightarrow}} \; CH_3 - \overset{\overset{\displaystyle CH_3}{|}}{CH} - CH_2 - CH = OH^+
$$

2-Methylbutanal

$$
\Updownarrow
$$

$$
CH_3 - \overset{\overset{\displaystyle CH_3}{|}}{CH} - CH = \overset{\overset{\displaystyle H}{|}}{C} - OH + H^+
$$

$$
\Big\downarrow \;-\;H_2O
$$

$$
CH_2 = \overset{\overset{\displaystyle CH_3}{|}}{C} - CH = CH_2
$$

Isoprene

Reactions of Acids and Esters

Conversion to hydrocarbons

The conversion of low-molecular-weight carboxylic acids, such as acetic acid, to hydrocarbons over HZSM-5 is accompanied by rapid catalyst deactivation and yields less hydrocarbon than that of methanol, because of their low effective hydrogen content.* However, total conversion of acetic acid can be maintained indefinitely provided that the catalyst residence time is short and periodic catalyst regeneration is employed.

It is interesting that decarboxylation reaction takes place extensively over ZSM-5. As a result of the rejection of oxygen as CO_x, a substantial amount of hydrocarbon is formed.

Addition of methanol to acetic acid decreases the rate of catalyst deactivation. More interestingly, it shifts the oxygen rejection reaction from decarboxylation to dehydration (Figure 4.22). As a result, a synergistic effect was observed (Chang et al., 1983) in increasing the yield of total hydrocarbons and C_5^+ liquid. The C_5^+ liquid from the acid/alcohol mixtures is highly aromatic (H/C ratio of about 1.3) compared to a H/C ratio of about 2 for the liquid product from methanol.

The behavior of esters is similar to that of acid/alcohol mixtures (Table 4.30).

High-molecular-weight fatty acids and triglycerides, on the other hand, have higher effective hydrogen content than acetic acid and are more hydrocarbon-like. Weisz et al. (1979) showed that natural products such as corn oil, peanut oil, castor oil, and

---

*For compounds containing oxygen, their effective hydrogen content or H/C ratio can be expressed as follows:

$$(H/C)_{eff} = \frac{H - 2(O \times w)}{C - O \times cm - O \times cd \times 0.5}$$

where H, C, O are the number of hydrogen, carbon and oxygen atoms in the empirical formula of the feed respectively, and w, cm, and cd are the fractions of the oxygen rejected as water, carbon monoxide, or carbon dioxide, respectively. Assuming all the oxygen to be rejected as water, then

$$(H/C)_{eff} = \frac{H - 2 \times O}{C}$$

For example, acetic acid has an effective H/C ratio of zero (Haag et al., 1980.)

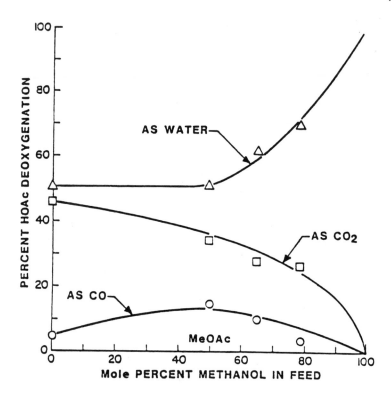

Figure 4.22   Effect of methanol on acetic acid deoxygenation.
(From Chang et al., 1983.)

jojoba oil can be converted over ZSM-5 to a mixture of paraffins,
olefins and aromatics, with a product distribution similar to
that from methanol (Figure 4.23).

Fries rearrangement
The intramolecular transformation of phenyl acetate into hydroxy-
acetophenones, known as the Fries rearrangement, was studied
by Pouilloux et al. (1987) over acid catalysts at 400°C. Compared to large pore zeolites, such as HY, the HZSM-5 catalyst
was more stable and remained active for hours. The para-to-
ortho ratio of hydroxyacetophenone produced over HZSM-5 was

Table 4.30   Conversion of Ester and Acid/Alcohol Mixture
over HZSM-5

410°C, 1 atm, 1.0-1.1 WHSV, 20 min reaction intervals		
	Methyl acetate	1.9/1 (molar) Methanol/ acetic acid
Total conversion	89.4	>91
Conversion to non-oxygenates	86.1	90.4
Products (wt % of charge)		
CO	6.2	2.1
CO_2	17.6	9.4
H_2O	21.5	45.3
Oxygenates	13.9	9.6
C_4^- Hydrocarbon gas	6.0	7.9
C_5^+ Liquid hydrocarbon	32.1	24.9
Total hydrocarbons	38.1	32.8
Coke	2.7	0.8
Wt %'s of hydrocarbon		
Methane + Ethane	5.6	7.2
Propane	0.7	0.4
Propene	6.7	13.8
Isobutane	0.3	0.4
n-Butane	0.0	0.1
Butenes	1.4	1.6
Subtotal	14.7	23.5
C_5^+ (gasoline)	78.7	74.1
Coke	6.6	2.4
H/C of C_5^+	1.3	~1.4

Source: Chang et al. (1983).

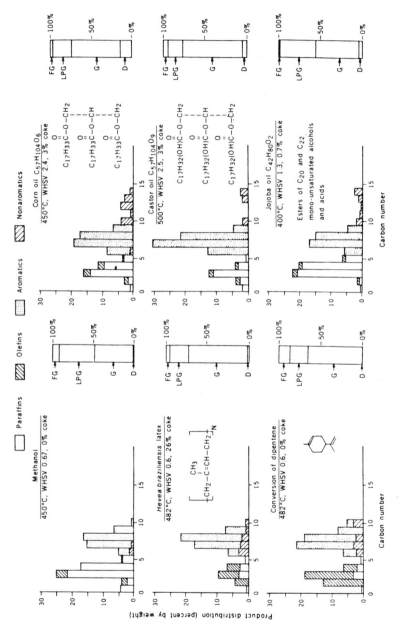

Figure 4.23 Product spectra from catalytic conversion of methanol, and of various biomass constituents. The abbreviations *FG, LPG, G* and *D* are for fuel gas, liquid petroleum gas, gasoline and light distillate respectively; *WHSV* is for weight-hourly space velocity. (From Weisz et al., 1979.)

about 10 times higher than that over nonshape selective catalysts.

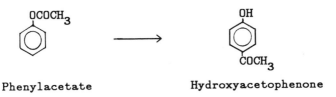

Phenylacetate   Hydroxyacetophenone

Carbohydrates

When carbohydrates are thermally decomposed, water molecules are driven off, leaving a carbon residue; thus removal of oxygen in the form of water molecules does not produce hydrocarbons. Oxygen must be removed as carbon monoxide or carbon dioxide in order to enrich product hydrogen.

With short catalyst residence time and frequent regeneration, hydrocarbons are produced directly from carbohydrates over HZSM-5. However, the hydrocarbon yield appears to decrease with the increasing molecular size of the sugar molecule, suggesting that the reactions leading to hydrocarbon formation depend on getting the sugar molecules into the intracrystalline pores of the zeolite (Table 4.31).

Figure 4.24 reveals that between 9% and 21.8% of the carbon in the carbohydrate fed to the reactor is converted to hydrocarbons.

Phenols

Isomerization

Similar to the isomerization of alkylbenzenes, the isomerization of substituted phenols having similar molecular sizes as that of xylenes, such as mono-alkylphenols, have also been reported.

Keim et al. (1981) showed that the isomerization of cresols proceeds over ZSM-5, ZSM-11, and ZSM-12.

The isomerization reaction can be carried out readily at 380-420°C. Selectivity in excess of 95% is achieved. Compared

Table 4.31   Direct Conversion of Carbohydrates at 510°C

	50% Xylose solution	50% Glucose solution	70% Starch solution	50% Sucrose solution
WHSV	1.9	2.1	2.1	2.1
Product, wt %				
Hydrocarbons	10.0	8.1	8.9	4.4
CO	33.3	18.9	16.8	32.8
CO_2	3.7	3.7	1.5	8.6
Coke	16.8	24.9	30.4	23.8
H_2O	36.2	44.4	42.4	33.4

Source: Chen et al. (1986).

to nonshape selective catalysts, medium pore zeolites inhibit the undesirable disproportionation reactions that lead to the formation of phenol and higher alkylated products. Typical experimental data on the isomerization of *o*-cresol are shown in Table 4.32.

The equation for isomerization of cresols follows:

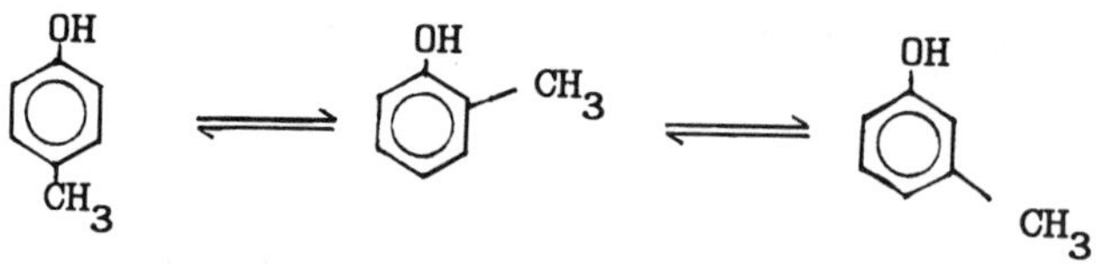

Cresols may also be prepared by the catalytic methylation of phenol. The electrophilic substitution reaction is strongly ortho-para-directing. The *m*-isomer can be made by subsequent isomerization.

Synthesis of long chain alkylphenols
As discussed earlier regarding the alkylation of benzene with long chain olefins over ZSM-12, Young (1981b) also found that this

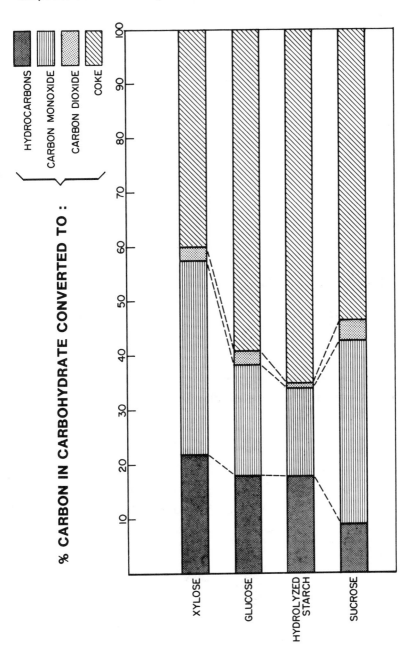

Figure 4.24 Product distribution of carbohydrates at 510 C, 2 WHSV (weight hourly space velocity), 1 atm. (From Chen et al., 1986.)

Table 4.32   Isomerization of o-Cresol

	ZSM-5	ZSM-12
Temperature, °C	380	420
LHSV	2.8	2.5
Pressure, atm	60	60
Product, wt %		
Phenol	1.5	3.9
o-Cresol	59.0	53.4
m-Cresol	28.3	27.7
p-Cresol	10.1	11.1
Xylenol	1.1	3.7
Trimethylphenols	0.0	0.2

Source: Keim et al. (1981).

zeolite catalyzes the preferential formation of para-mono-alkyl-phenols when phenol is alkylated with long chain olefins; their data are shown in Table 4.13.

F. Reactions of Nitrogen-Containing Compounds

Aromatic Amines

Reaction between phenols, alicyclic alcohols, or ketones, such as cyclohexanol or cyclohexanone and ammonia or alkylamines, over ZSM-5 yields selected aromatic amines such as aniline and p-toluidine (Chang and Lang, 1983; 1984). Such bulky products as diphenylamine and carbazole are apparently excluded.

Aromatic amines can also undergo isomerization reactions over the medium pore zeolites. Isomerization of toluidines proceeds readily over HZSM-5 with little or no aging (Eichler et al., 1985). Weigert (1986) showed that toluidines are isomerized over ZSM-5 at 300°C to 400°C and atmospheric pressure to an equilibrium mixture (Figure 4.25).

Eichler et al. (1986) showed that 2-ethyl aniline is isomerized to 3-ethyl aniline with no olefinic by-products. Interestingly, consistent with the critical dimension of the isomers, the isomerization of dimethylanilines (Weigert, 1986) occurs only with the interconversion among three isomers (2,4-, 2,5- and 3,4-) while the other three isomers (2,3-, 2,6- and 3,5-) are neither formed nor reactive over ZSM-5.

Toluidine Isomerization on HZSM5

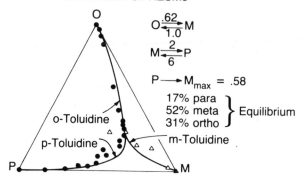

Figure 4.25   Toluidine isomerization on HZSM-5. (From Weigert, 1986.)

Pyridines

Mixtures of pyridine and alkyl pyridines can be synthesized with excellent yields by reacting carbonyl compound, such as acetaldehyde, formaldehyde, and acetone, with ammonia (Chang and Lang, 1980) over HZSM-5. The reaction takes place at 425°C and atmospheric pressure. The selectivity for pyridine production can be enhanced when the reaction is carried out in a fluid bed with short residence times (Feitler et al., 1987). Reacting ethanol with ammonia to form mixed pyridine bases over medium pore zeolites was reported by van der Gaag et al. (1986).

$$6CH_3 - C{<}^O_H + 2NH_3 \longrightarrow \quad + \quad + 6H_2O + 2H_2$$

α – picoline γ – picoline

$$4CH_3 - C{<}^O_H + 3HC{<}^O_H + 2NH_3 \longrightarrow \quad + \quad + 7H_2O + H_2$$

pyridine β–picoline

Compared to large pore zeolites, such as mordenite, the HZSM-5 catalyst was more stable and produced much less poly-alkyl pyridines and heavy products, as expected (Table 4.33).

α-Picoline can also be selectively produced from aniline or phenol with ammonia over the ZSM-5 catalyst (Chang and Perkins, 1983). By increasing the ratio of ammonia to phenols fed, the reaction can be directed toward the selective production of α-picoline without making either the β- or the γ-isomers.

Le Blanc et al. (1986) reported that 2-amino-6-alkyl pyridine can be selectively produced by isomerizing 1,3-diamino-benzene or reacting, in the presence of ammonia, 3-aminophenol or 1,3-dihydroxybenzene over a medium pore zeolite. The reaction, specific to the meta isomer, is carried out at about 400°C under pressure.

Diaminophenol   2-Amino-6-methyl
                                     pyridine

Nitriles

Chang et al. (1980) also found that by reacting a mixture of formaldehyde, alcohol, and ammonia or alkylamines over ZSM-5, alkylnitriles and alkylaminonitriles are produced. Ammoxidation of toluene to benzonitrile over a copper-exchanged HZSM-5 was reported by Oudejans et al. (1984). These reactions are given:

Alkylnitriles, alkylaminitriles

$$HCHO + CH_3OH + NH_3 \longrightarrow CH_3CN + C_2H_5CN$$
$$CH_3-NH-CH_2-CN + (CH_3)_2-N-CH_2-CN$$
$$+ CH_3-NH-CH_2-OH_2-CN + H_2O$$

Table 4.33   Reaction of Acetaldehyde and Ammonia

Catalyst	HZSM-5		H-Mordenite	
Time on stream, hrs	0.5	3.0	0.5	3.0
Acetaldehyde, % conversion	80	60	65	4
Products, wt %				
Light products	4.0	4.5	9.5	9.3
Pyridine	44.9	45.3	28.5	9.0
Picolines	28.1	24.5	25.2	13.9
Lutidines	17.0	19.6	27.3	33.8
Heavy products	5.9	6.1	9.5	34.0

Source: Chang and Lang (1980).

Acetonitriles

$$3 \ HCHO + NH_3 \longrightarrow CH_3CN + H_2O$$

Benzonitrile

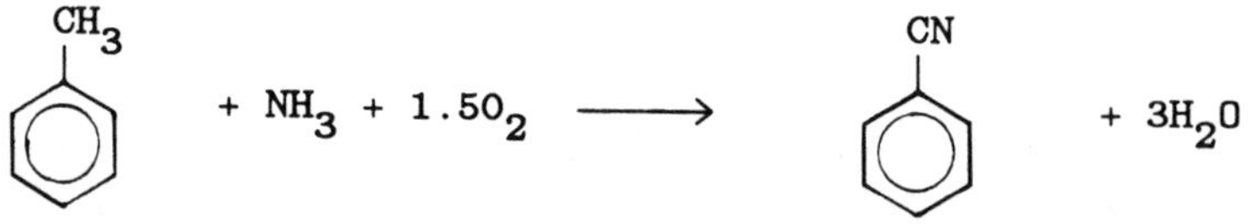

Caprolactam

The Beckmann rearrangement of cyclohexanone oxime to ε-caprolactam when carried out over an acid catalyst such as zeolite Y (Landis and Venuto, 1966; Aucejo et al., 1986) is accompanied by rapid coking and loss of selectivity. Bell and Chang (1982) showed that with medium pore zeolites, the catalysts remain active for hours. Compared to using H_2SO_4, the solid acid catalyst eliminates the by-product, ammonium sulfate.

Cyclohexanone oxime   Caprolactam

Cyanamide

The synthesis of cyanamide from carbon dioxide or urea and ammonia is usually accompanied by the formation of undesirable by-products of oligomerization reactions, such as melamine, a cyanamide trimer. Van Hardeveld et al. (1986) reported that these secondary reactions can be avoided by using a zeolite catalyst with pores openings less than 8 Å. The reaction is carried out at about 400°C with isocyanic acid or urea, in the presence of ammonia.

$$2 \; HN=C=O \longrightarrow H_2N-CN \; + \; CO_2$$

Isocyanic   Cyanamide
   acid

G. Reactions of Other Non-Hydrocarbons

Conversion to Hydrocarbons

Similarly to methanol and many other oxygen containing compounds, many compounds containing halogen, sulfur or nitrogen can be converted to hydrocarbons readily over the medium pore zeolites (Chang et al., 1975a). Butter et al. (1975) reported the conversion of aliphatic mercaptans, aliphatic sulfides, aliphatic halides, and aliphatic amines to a mixture of gaseous and liquid hydrocarbons. Instead of producing water as the major byproduct, hydrogen is rejected alone with the halogen, sulfur, or nitrogen in the molecule.

As in the case of compounds containing oxygen, the hydrocarbon yields and the rate of catalyst deactivation depend on the effective hydrogen content of the feed, defined as:

$$(H/C)_{\text{eff}} = \frac{H - 2 \times (O + S + X) - 3 \times N)}{C}$$

where H, C, S, X, and N are the number of hydrogen, carbon, oxygen, sulfur, halogen, and nitrogen atoms in the empirical formula of the feed respectively, assuming all the oxygen to be rejected as water.

Chang and Perkins (1987) also showed that low effective hydrogen feeds, such as chloroform, $CHCl_3$ can react with methane over ZSM-5 to form methylchloride, which in turn can be converted to higher-molecular-weight hydrocarbons over the same catalyst.

Isomerization

The isomerization activity of medium pore zeolites has also been studied over a variety of non-hydrocarbon aromatics and hetero-cyclic compounds, including dichlorotoluenes (Litterer, 1985; Toray Ind., 1985), halogenated thiophenes (Eichler and Leupold, 1986) and halogenated phenols (Baltes and Leupold, 1986). The latter reported that the isomerization of 2-chlorophenol and/or 4-chlorophenol to an equilibrium mixture of monochlorophenols takes place readily over HZSM-5.

Isomerization of chlorophenols

H. Acid-Catalyzed Hydrogenation Activity

The controversial question of whether or not acid sites in zeolites catalyze hydrogenation reactions (Minachev et al., 1968; Chen and Garwood, 1977) has received some renewed attention in more recent years. Dadalla et al. (1983) reconfirmed the earlier studies on mordenite (Minachev et al., 1971; 1972) that concluded that Zeolon-500, a synthetic mordenite, is active for the hydrogenation of ethene. Chang and Socha (1984), in their study on the conversion of synthesis gas to hydrocarbons, found that HZSM-5 has small but significant activity for hydrogenating carbon monoxide to light saturated hydrocarbons. The role of molecular hydrogen in acid catalyzed reactions was confirmed by Haag and Dessau (1984). They showed that the restricted pore geometry of the medium pore zeolites promotes a monomolecular paraffin cracking reaction that produces molecular hydrogen. More recently, Sano et al. (1987) also demonstrated the hydrogenation of benzene over HZSM-5.

REFERENCES

Alberty, R. A., J. Phys. Chem. *87*, 4999 (1983).

Asahi Chem., Japan Pat. 59-219241, Dec. 9, 1984.

Aucejo, A., M. C. Burguet, A. Corma, and V. Fornes, Appl. Catal. *22*, 187 (1986).

Baltes, H. and E. I. Leupold, U.S. Pat. 4,568,777, Feb. 4, 1986.

Becker, K. A., H. G. Karge, and W. D. Streubel, J. Catal. *28*, 403 (1973).

Bell, W. K. and C. D. Chang, U.S. Pat. 4,359,421, Nov. 16, 1982.

Bonacci, J. C. and R. P. Billings, U.S. Pat. 3,948,758, Apr. 6, 1976.

Brennan, J. A. and R. A. Morrison, U.S. Pat. 3,945,913, Mar. 23, 1976.

Brennan, J. A., W. E. Garwood, S. Yurchak, and W. Lee, Proc. Int. Sem. on Alternate Fuels, Liege, Belgium, May 1981, A. Germain, ed., p. 19.1 (1981).

Brown, H. C. and C. R. Smoot, J. Am. Chem. Soc. *78*, 6255 (1956).

Butter, S. A., A. T. Jurewicz, and W. W. Kaeding, U.S. Pat. 3,894,107, Jul. 8, 1975.

Butter, S. A., U.S. Pat. 3,979,472, Sept. 7, 1976.

Caesar, P. D. and R. A. Morrison, U.S. Pat. 4,083,889, Apr. 11, 1978.

Ceckiewicz, S., J. Chem. Soc. Faraday Trans. I 77, 269 (1981).

Chang, C. D., W. H. Lang, and A. J. Silvestri, U.S. Pat. 3,894,104, Jul. 8, 1975a.

Chang, C. D., A. J. Silvestri, and R. L. Smith, U.S. Pat. 3,894,105, Jul. 8, 1975b.

Chang, C. D. and A. J. Silvestri, J. Catal. *47*, 249 (1977).

Chang, C. D., W. H. Lang, and A. J. Silvestri, U.S. Pat. 4,062,905, Dec. 13, 1977.

Chang, C. D., W. H. Lang, and R. L. Smith, J. Catal. *56*, 169 (1979).

Chang, C. D. and W. H. Lang, U.S. Pat. 4,220,783, Sept. 2, 1980.

Chang, C. D. and N. J. Morgan, U.S. Pat. 4,214,107, Jul. 22, 1980.

Chang, C. D., W. H. Lang, and R. B. LaPierre, U.S. Pat. 4,231,955, Nov. 4, 1980.

Chang, C. D., W. H. Lang, and W. K. Bell, *Catalysis of Organic Reactions*, W. R. Moser, ed., Marcel Dekker, New York, p. 73, 1981.

Chang, C. D., Catal. Rev.-Sci. Eng. *25*, 1 (1983); *Hydrocarbons from Methanol*, Marcel Dekker, New York, 1983.

Chang, C. D. and W. H. Lang, U.S. Pat. 4,380,669, Apr. 19, 1983.

Chang, C. D. and P. D. Perkins, U.S. Pats. 4,388,461, June 14, 1983; 4,395,554, July 26, 1983.

Chang, C. D., N.Y. Chen, L. R. Koenig, and D. E. Walsh, Am. Chem. Soc. Div. Fuel Chem. Preprints *28*(2), 146 (1983).

Chang, C. D. and W. H. Lang, U.S. Pat. 4,434,299, Feb. 28, 1984.

Chang, C. D. and R. F. Socha, U.S. Pat. 4,472,535, Sept. 18, 1984.

Chang, C. D., C. T.-W. Chu, and R. F. Socha, J. Catal. *86*, 289 (1984).

Chang, C. D. and S. D. Hellring, U.S. Pat. 4,620,044, Oct. 28, 1986.

Chang, C. D. and P. D. Perkins, U.S. Pat. 4,654,449, Mar. 31, 1987.

Chang, C. L., "Dehydration of Ethanol over ZSM-5," doctoral dissertation, Michigan State University, 1985; Diss. Abstr. Int. B *47*(2) 711 (1986).

Chen, N.Y., J. Maziuk, A. B. Schwartz, and P. B. Weisz, Oil Gas
    J. 66(47), 154 (1968).
Chen, N.Y., U.S. Pat. 4,002,697, Jan. 11, 1977.
Chen, N.Y. and W. E. Garwood, J. Catal. 50, 252 (1977).
Chen, N.Y. and W. J. Reagan, U.S. Pat. 4,049,735, Sept. 20,
    1977.
Chen, N.Y., R. L. Gorring, H. R. Ireland, and T. R. Stein, Oil
    Gas J. 75(23), 165 (1977).
Chen, N.Y., U.S. Pat. 4,100,215, July 11, 1978.
Chen, N.Y. and W. E. Garwood, J. Catal. 53, 284 (1978a).
Chen, N.Y. and W. E. Garwood, J. Catal. 52, 453 (1978b).
Chen, N.Y. and W. J. Reagan, J. Catal. 59, 123 (1979a).
Chen, N.Y. and W. J. Reagan, U.S. Pat. 4,148,835, Apr. 10, 1979b.
Chen, N.Y., W. E. Garwood, W. O. Haag, and A. B. Schwartz, "Shape
    Selective Hydrocarbon Catalysis Over Synthetic Zeolite
    ZSM-5," paper presented at Symp. Advan. Catal. Chem. I,
    Snowbird, Utah, Oct. 3, 1979(a).
Chen, N.Y., W. W. Kaeding, and F. G. Dwyer, J. Am. Chem. Soc.
    101, 6783 (1979b).
Chen, N.Y. and W. J. Reagan, U.S. Pat. 4,278,565, July 14,
    1981.
Chen, N.Y., "Ethanol Fuel from Biomass," paper presented at
    the Tri-State Catalysis Club, Lexington, Ky., Oct. 20, 1982.
Chen, N.Y., Chemtech 13, 488 (1983).
Chen, N.Y. and T. Y. Yan, Ind. Eng. Chem., Process Design
    Devel. 25, 151 (1986).
Chen, N.Y. and W. E. Garwood, Ind. Eng. Chem., Prod. Res.
    Devel. 25, 641 (1986).
Chen, N.Y., T. F. Degnan, and L. R. Koenig, Chemtech. 16, 506
    (1986).
Chu, C. T.-W. and C. D. Chang, J. Catal. 86, 297 (1984).
Chu, P. C. and G. H. Kuehl, Ind. Eng. Chem. Res. 26, 365
    (1987).
Cortes, A. and A. Corma, J. Catal. 51, 338 (1978).
Dadalla, A. M., T. C. Chan, and R. G. Anthony, Int. J. Chem.
    Kinet. 15, 759 (1983).
Dejaifve, P., J. C. Vedrine, and E. G. Derouane, J. Catal. 63,
    331 (1980).
Derouane, E. G. and J. C. Vedrine, J. Mol. Catal. 8, 479 (1980).

Derouane, E. G., Stud. Surf. Sci. Catal. *20*, 221 (1985).

Dessau, R. M. and R. B. LaPierre, J. Catal. *78*, 136 (1982).

Dessau, R. M., J. Catal. *89*, 520 (1984).

Donnelly, S. P. and J. R. Green, "Recent MDDW Process Improvements," paper presented at the Symp. Jap. Petrol. Inst., Tokyo, Oct. 2-3, 1980.

Dwyer, F. G. and D. J. Klocke, U.S. Pat. 4,049,737, Sept. 20, 1977.

Eichler, K., E. Leupold, H. J. Arpe, and H. Baltes, German Pat. 3420707, Dec. 5, 1985.

Eichler, K. H. J. Arpe, and E. I. Leupold, European Pat. 175228, Mar. 26, 1986.

Eichler, K. and E. Leupold, U.S. Pat. 4,604,470, Aug. 5, 1986.

Espinoza, R. F. and W. G. B. Mandersloot, J. Mol. Catal. *24*, 127 (1984).

Feitler, D., U.S. Pat. 4,612,387, Sept. 19, 1986.

Feitler, D., D. W. Schimming, and H. Wetstein, U.S. Pat. 4,675,410, June 23, 1987.

Fowles, P. E. and T. Y. Yan, U.S. Pats. 4,524,227, 4,524,228, 4,524,231, June 18, 1985.

Galya, L. G., M. L. Occelli, J. T. Hsu, and D. C. Young, J. Mol. Catal. *32*, 391 (1985).

Garwood, W. E. and N. Y. Chen, "Octane Boosting Potential of Catalytic Processing of Reformate over Shape Selective Zeolite," paper presented at the Am. Chem. Soc. Mtg., Houston, Mar. 23-28, 1980; Div. Petrol. Chem. Preprint, *25*(1), 84 (1980).

Garwood, W. E., Am. Chem. Soc. Symp. Ser. *218*, 383 (1983).

Givens, E. N., C. J. Plank, and E. J. Rosinski, U.S. Pat. 4,079,095 and 4,079,096, Mar. 14, 1978.

Gorring, R. L., J. Catal. *31*, 13 (1973).

Gould, R. M., A. A. Avidan, J. L. Soto, C. D. Chang, and R. F. Socha, "Scale-Up of a Fluid-Bed Process for Production of Light Olefins from Methanol," paper presented at the AIChE Nat. Mtg., New Orleans, Apr. 6-10, 1986.

Haag, W. O. and F. G. Dwyer, "Aromatic Processing with Intermediate Pore Size Zeolite Catalysts," paper presented at the AIChE 8th Natl. Mtg., Boston, Aug. 19, 1979.

Haag, W. O. and D. H. Olson, U.S. Pat. 4,117,026, Sept. 26, 1978.

Haag, W. O., P. G. Rodewald, and P. B. Weisz, "Catalytic Production of Aromatics and Olefins from Plant Materials," paper presented at the Am. Chem. Soc. Nat. Mtg., San Francisco, Aug. 24 -29, 1980.

Haag, W. O., R. M. Lago, and P. B. Weisz, Faraday Disc. 72, 317 (1982a).

Haag, W. O., R. M. Lago, and P. G. Rodewald, J. Mol. Catal. 17, 161 (1982b).

Haag, W. O. and R. M. Lago, U.S. Pat. 4,374,296, Feb. 15, 1983a.

Haag, W. O. and R. M. Lago, U.S. Pat. 4,418,235, Nov. 29, 1983b.

Haag, W. O., Proc. 6th Int. Zeolite Conf., July 10–15, 1983, Reno, Nev., D. Olson and A. Basio, eds., Butterworths, Surrey, U.K., p. 466, 1984.

Haag, W. O., R. M. Lago, and P. B. Weisz, Nature 309, 589 (1984).

Haag, W. O. and R. M. Dessau, Proc. 8th Int. Congr. Catal. 2, 305 (1984).

Hagen, G. P., U.S. Pat. 4,433,174, Feb. 21, 1984.

Heinemann, H., Catal. Rev.-Sci. Eng. 15, 53 (1977).

Hoelderich, W., F. Merger, W. D. Mross, and R. Fischer, European Pat. 162387, Nov. 27, 1985a.

Hoelderich, W., F. Merger, W. D. Mross, and G. Fouquet, U.S. Pat. 4,560,822, Dec. 24, 1985b.

Hoelderich, W., Proc. 7th Int. Zeolite Conf., Y. Murakami, A. Iijima, and J. W. Ward, eds., Kodadansha/Elsevier, p. 827 (1986).

Huang, T. J. and W. O. Haag, U.S. Pat. 4,339,606, July 13, 1982.

Ihara Chem., Japan Pat. 60-048943, Mar. 16, 1985a.

Ihara Chem., European Pat. 154236, Sept. 11, 1985b.

Inui, T., T. Ishihara, and Y. Takegami, J. Chem. Soc., Chem. Commun., 936 (1981).

Ishida, H. and H. Nakajima, U.S. Pat. 4,613,717, Sept. 23, 1986.

Jacobs, P. A., J. B. Uytterhoeven, M. Steyns, G. Froment, and J. Weitkamp, Proc. 5th Int. Conf. Zeolites, L. V. Rees, ed., Heyden, London, p. 607, 1980.

Jacobs, P. A., J. A. Martens, J. Weitkamp, and H. K. Beyer, Faraday Disc. Chem. Soc. 72, 353 (1982).

Kaeding, W. W. and L. B. Young, U.S. Pat. 4,094,921, June 13, 1978.

Kaeding, W. W. and S. A. Butter, J. Catal. 61, 155 (1980).

Kaeding, W. W., M. M. Wu, L. B. Young, and G. T. Burress, U.S. Pat. 4,197,413, Apr. 8, 1980.

Kaeding, W. W., C. Chu, L. B. Young, B. Weinstein, and S. A. Butter, J. Catal. *67*, 159 (1981a).

Kaeding, W. W., C. Chu, L. B. Young, and S. A. Butter, J. Catal. *69*, 392 (1981b).

Kaeding, W. W., L. B. Young, and A. G. Prapas, Chemtech *12*, 556 (1982).

Kaeding, W. W., J. Catal. *95*, 512 (1985).

Kaiser, S. W., U.S. Pat. 4,499,327, Feb. 12, 1985; U.S. Pat. 4,524,234, June 18, 1985.

Karge, H. G., J. Ladebeck, Z. Sarbak, and K. Hatada, Zeolites *2*, 94 (1982).

Karge, H. G., Y. Wada, J. Weitkamp, S. Ernst, U. Girrbach, and H. K. Beyer, Stud. Surf. Sci. Catal. *19*, 101 (1984).

Keim, K., R. Kiauk, and E. Meisenburg, U.S. Pat. 4,283,571, Aug. 11, 1981.

Landis, P. S. and P. B. Venuto, J. Catal. *6*, 245 (1966).

Le Blanc, H., L. Puppe, and K. Wedemeyer, U.S. Patent 4,628,097, Dec. 9, 1986.

Linstid, H. C. and G. S. Koermer, U.S. Pat. 4,537,995, Aug. 27, 1985.

Litterer, H., German Pat. 3334673, Apr. 11, 1985.

Long, G. N., R. J. Pellet, and J. A. Rabo, U.S. Pat. 4,528,414, July 9, 1985.

Martens, J. A., "Mechanism of Isomerization and Hydrocracking of Long-Chain Paraffins on Large Pore and Shape Selective Zeolites," Ph.D. thesis, Catholic University Leuven, Belgium, Dec. 1985.

Minachev, Kh. M., O. K. Shchudina, M. A. Markov, and R. V. Dimitriev, Neftekhimiya *8*, 37 (1968).

Minachev, Kh. M., V. I. Garanin, T. A. Isakova, V. V. Kharlamov, and V. Bogomolov, Advan. Chem. Ser. *102*, 441 (1971).

Minachev, Kh. M., V. I. Garanin, V. V. Kharlamov, and T. A. Isakova, Kinet. Katal. *13*, 1101 (1972).

Nagy, J. B., J. Mol. Catal. *5*, 393 (1979).

Olson, D. H., G. T. Kokotailo, S. L. Lawton, and W. H. Meier, J. Phys. Chem. *85*, 2238 (1981).

Olson, D. H. and W. O. Haag, Am. Chem. Soc. Symp. Ser. *248*, 275 (1984).

Ono, Y. and T. Mori, J. Chem. Soc. Faraday Trans. I 77, 2209 (1981).

Onodera, T., T. Sakai, Y. Yamasaki, and K. Sumitani, U.S. Pat. 4,320,242, Mar. 16, 1982.

Oudejans, J. C., F. J. van der Gaag, and H. van Bekkum, Proc. 6th Int. Zeolite Conf., July 10-15, 1983, Reno, Nev., D. Olson and A. Bisio, eds., Butterworths, Surrey, U.K., p. 536, 1984.

Pecci, G. and T. Floris, Hydrocarbon Process 56(12), 98 (1977).

Pellet, R. J., J. A. Rabo, G. N. Long, and P. K. Coughlin, "Reactions of C_8 Aromatics Catalyzed by Aluminophosphate Based Molecular Sieves," Paper No. B-2, 10th N. Am. Mtg. Catal. Soc., San Diego, May 17-22, 1987.

Pouilloux, Y., N. S. Gnep, P. Magnoux, and G. Perot, J. Mol. Catal. 40, 231 (1987).

Poutsma, M. L., "Zeolite Chemistry and Catalysis," J. A. Rabo, ed., Am. Chem. Soc. Monogr. 171, 437 (1976).

Quann, R. J., L. A. Green, S. A. Tabak, and F. J. Krambeck, "Chemistry of Olefin Oligomerization over ZSM-5 Catalyst," paper presented at the AIChE Nat. Mtg., New Orleans, April 6-10, 1986.

Ratnasamy, P., G. P. Babu, A. J. Chandwadkar, and S. B. Kulkarni, Zeolites 6, 98 (1986).

Rodewald, P. G., U.S. Pat. 4,100,219, July 11, 1978; U.S. Pat. 4,145,315, Mar. 20, 1979.

Sano, T., O. Kiyomi, H. Hagiwara, H. Takaya, H. Shoji, and K. Matsuzaki, J. Mol. Catal. 40, 113 (1987).

Sato, H., U.S. Pat. 4,465,852, Aug. 14, 1984.

Sato, H., N. Ishii, and S. Nakamura, U.S. Pat. 4,499,321, Feb. 12, 1985.

Sato, A., S. Kawakami, H. Dohi, and K. Endo, U.S. Pat. 4,568,793, Feb. 4, 1986.

Singh, B. B. and R. G. Anthony, Prepr. Can. Symp. Catal. 6, 113 (1979).

Singh, B. B., F. N. Lin, and R. G. Anthony, Chem. Eng. Commun. 4, 749 (1980).

Smith, K. W., W. C. Starr, and N. Y. Chen, Oil Gas J. 78(21), 75 (1980).

Stock, L. M. and H. C. Brown, J. Am. Chem. Soc. 81, 3323 (1959).

Sullivan, R. F., C. J. Egan, G. E. Langlois, and R. P. Sieg, J. Am. Chem. Soc. 83, 1156 (1961).

Tabak, S. A., F. J. Krambeck, and W. E. Garwood, "Conversion of Propylene and Butylene Over ZSM-5 Catalyst," paper presented at the AIChE Mtg., San Francisco, Nov. 25-30, 1984; AIChE J. 32, 1526 (1986).

Toray Ind., Japan Pat. 60-042340, Mar. 6, 1985.

Toray Ind., Japan Pat. 61-050933, Mar. 13, 1986.

Van den Berg, J. P., J. P. Wolthuizen, and J. H. C. van Hooff, Proc. 5th Intn. Conf. Zeolites, L. V. Rees, ed., Heyden, London, p. 649 (1980).

Van den Berg, J. P., J. P. Wolthuizen, A. D. H. Clague, G. Hays, R. Huis, and J. H. C. van Hooff, J. Catal. *80*, 130 (1983); J. Catal. *80*, 139 (1983).

Van der Gaag, F. J., F. Louter, J. C. Oudejans, and H. van Bekkum, Appl. Catal. *26*, 191 (1986).

Van Hardeveld, R., T. J. van de Mond, and F. H. Vandenbooren, U.S. Pat. 4,625,061, Nov. 25, 1986.

Venuto, P. B., J. Org. Chem. *32*, 1272 (1967).

Venuto, P. B. and P. S. Landis, Advan. Catal. *18*, 259 (1968).

Wei, J., J. Catal. *76*, 433 (1982).

Weigert, F. J., U.S. Pat. 4,593,124, June 3, 1986.

Weisz, P. B., W. O. Haag, and P. G. Rodewald, Science *206*, 57 (1979).

Weitkamp, J., P. A. Jacobs, and J. A. Martens, Appl. Catal. *8*, 123 (1983).

Wise, J. J., U.S. Pat. 3,251,897, May 17, 1966.

Wu, M. M. and W. W. Kaeding, J. Catal. *88*, 478 (1984).

Wunder, F. A. and E. I. Leupold, Angew. Chem. Int. Ed. Engl. *19*(2), 126 (1980).

Young, L. B., U.S. Pat. 4,301,317, Nov. 17, 1981(a).

Young, L. B., U.S. Pat. 4,283,573, Aug. 11, 1981(b).

Young, L. B., U.S. Pat. 4,365,084, Dec. 21, 1982.

Young, L. B., S. A. Butter, and W. W. Kaeding, J. Catal. *76*, 418 (1982).

Young, L. B., U.S. Pat. 4,371,714, Feb. 1, 1983.

5
Applications in Petroleum Processing

I. OCTANE BOOSTING

A. Reforming of Naphthas

Reforming of naphtha to high octane gasoline uses a $Pt-Al_2O_3$
(or bimetallic) catalyst under hydrogen pressure to carry out a
number of reactions, including converting 6-ring and 5-ring
naphthenes to aromatics via dehydrogenation and dehydroiso-
merization, and converting paraffins to higher octane isoparaffins
and aromatics via hydroisomerization and dehydrocyclization.
However, two limitations are inherent in this "reforming" technology. First, the product always contains an essentially
equilibrium distribution of paraffin isomers and hence a significant amount of low octane normal paraffins remains in the
product, limiting the product octane. Second, not only is the
paraffin dehydrocyclization efficiency for C_7^- paraffins quite
low (Ramage et al., 1987) (Figure 5.1), but there is also a
significant volume contraction when paraffins are converted to
aromatics, lowering the volumetric product yield.

   Thus, the reforming of light naphthas produces a low yield
of reformate. In the case of full range naphthas, high severity
reforming, about 98 R+O, is generally achieved by concentrating
the aromatics and hydrocracking the low octane component to
gases; hence this route also produces a low liquid yield.

   The addition of a small amount of low acidity ZSM-5 to the
reforming catalyst (Plank et al., 1979; 1981) selectively cracks
the low octane paraffinic molecules to propane and butanes, and
increases the octane boosting potential of the reforming catalyst.
For example, the data in Table 5.1 shows that under identical
operating conditions, the reforming catalyst containing 1 wt %

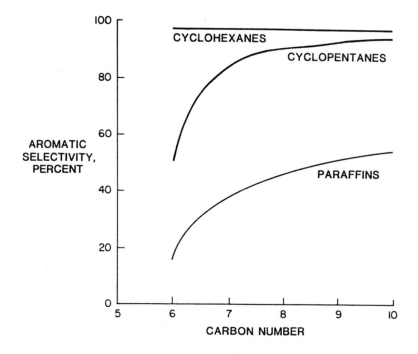

Figure 5.1   Component reforming efficiencies at 480°C, 15 atm. (From Ramage et al., 1987.)

of ZSM-5 increased the C_5^+ octane by 7.3 numbers over the base case. The comparison of these two catalysts at the same octane severity shows that the ZSM-5 containing reforming catalyst is about 20°C more active than the base case. The gain in propane and butanes also leads to a higher purity recycle gas, a key factor for stable operation.

B. Post-Reforming of Reformates

Selectoforming

In the mid-1960s, a shape selective post-reforming process, "Selectoforming," was introduced (Chen et al., 1968).

Table 5.1   Effect of ZSM-5 on Reforming Catalyst Performance

Feed: C_6-140°C midcontinent naphtha

Operating conditions:
Pressure: 14.6 atm, LHSV: 1.7,
hydrogen/hydrocarbon: 9.6 mol/mol

Catalyst	Base case	Base catalyst +1	wt % ZSM-5
Temperature, °C	485	485	465
$C_5{}^+$ Octane	94.5	101.8	96.4
$C_5{}^+$ Yield, vol %	83.4	69.7	74.4
Gas yields, wt %			
methane, ethane	1.8	3.0	1.6
propane	2.4	9.1	7.7
butanes	2.9	9.8	9.1

Source: Plank et al. (1979).

Selectoforming was the first commercial molecular shape selective catalytic process. It was designed to upgrade the octane rating of the reformate and produces LPG (primarily propane) as the major by-product. It is operated under hydrogen pressure and uses erionite, a 8-membered ring small pore zeolite, as the catalyst (see Chapter 2).

Based on the principle of size exclusion, the Selectoforming catalyst selectively converts the low octane n-paraffins in a reformate to propane. In addition to size exclusion of the feed molecules, the selective production of propane instead of iso-butane as the principal product of acid cracking represents another example of the size exclusion principle as applied to the product molecules.

To assure catalyst stability, a weak metal function, such as the sulfides of nickel, was incorporated into the catalyst to saturate olefins and prevent coke formation but not to hydrogenate the aromatics present in the feed (Chen and Garwood, 1968; Chen and Rosinski, 1971).

The acidity of the catalyst was generated by ion exchanging the zeolite with ammonium ions (Table 5.2). About 20–25% of the exchangeable sites are occupied by potassium ions situated inside the small cancrinite cages. This residual potassium may be removed by additional calcination and ammonium exchange steps. However, at below 450°C, the maximum operating temperature of the process, the rate of migration of potassium, which deactivates the catalyst, is very slow. Shown in Figure 5.2 is the commercial performance of a catalyst prepared without removing the residual potassium. The catalyst had a cycle life in excess of 1½ years.

Figure 5.3 shows a number of reactor configurations by which the Selectoforming process has been practiced commercially. Placing the catalyst in the bottom of the last reforming reactor is a simple and low cost approach, but it does not have the flexibility of using a separate reactor that can be operated at a different temperature and/or on a part-time basis when the market for propane is good.

The process generally operates at the reformer pressure. Reactor temperature may vary between 315°C and the temperature of the last reformer reactor depending on the reactor configuration. Recycle gas containing 60 to 80 mol % hydrogen may be used, while the gas-to-oil ratio is maintained at between 2 and 4.

The charge stocks can be unstabilized, debutanized, or depentanized reformates. Table 5.3 shows the yield shift after processing an unstabilized reformate. Case 1 is a situation in which the reformer throughput is increased and the low octane intermediate product from the reformer is processed over the Selectoforming reactor to the same final octane level and co-produce propane.

Case 2 shows the ability of the Selectoforming catalyst to boost the octane of the product beyond that of the reformer.

It is interesting to note that because the concentration of isoparaffins is higher in the Selectoformate than that of conventional

Table 5.2   Chemical Composition of Natural Erionite and Its H-Form

As received, wt %		Cation distribution equiv./Al		H-form cation distribution equiv./Al
SiO_2	65.9			
Al_2O_3	15.0			
K_2O	3.5	K	0.25	0.25
CaO	3.0	Ca	0.36	0.06
Na_2O	6.7	Na	0.73	0.01
MgO	0.7	Mg	0.14	0.05
Fe_2O_3	3.8	Fe	0.32	H+	0.63
Others	1.2		1.80	1.00
	100.0	SiO_2/Al_2O_3 = 7.5		SiO_2/Al_2O_3 = 7.4

Source: Chen and Garwood (1973a).

reformates at the same research octane, Selectoformates have a higher motor octane number, which is translated to higher road octane ratings (Chen et al., 1968; Burd and Majiuk, 1972).*

To produce propane, the Selectoforming process may be directly applied to light naphthas without reforming. Shown in Table 5.4 is the pilot plant result of Selectoforming a C_5^- 82°C light naphtha. The process selectively converts C_5^+ n-paraffins to propane and n-butane and raises the octane rating of the remaining liquid product. Although not in commercial practice,

---

*A research octane number and a motor octane number are two standard laboratory measures of a fuel's ability to resist knock during combustion in a spark ignition engine. These methods have been established by ASTM (1973) and are conducted in the laboratory by running the fuel in a single-cylinder test engine under prescribed operating conditions. The road octane number is an average of these two numbers (Benson, 1976).

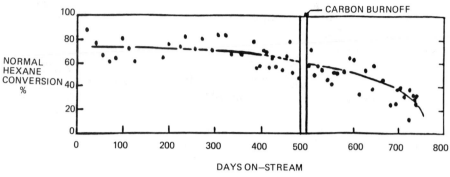

(a)

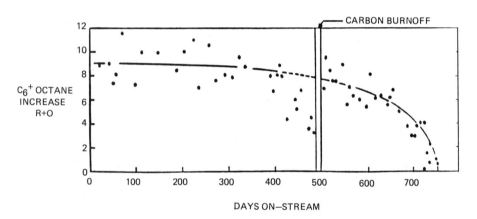

(b)

Figure 5.2 (a) Normal hexane conversion across selectoforming catalyst in commercial operation. (b) Research clear octane number increase across selectoforming catalyst in commercial operation. (From Roselius et al., 1973.)

SELECTOFORMING IN THIRD REACTOR OF REFORMER TRAIN (HIGH TEMPERATURE)

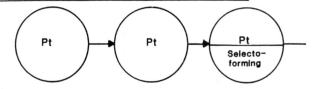

SELECTOFORMING IN THIRD REACTOR OF REFORMER TRAIN (LOW TEMPERATURE)

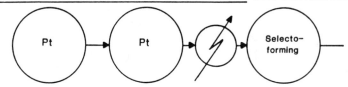

SELECTOFORMING IN SEPARATE REACTOR

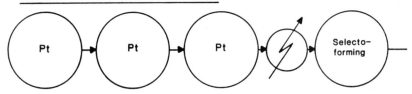

Figure 5.3   Catalyst fill configurations for reforming/selecto-forming process.

laboratory data indicate that under more severe operating conditions n-butane can be selectively converted to propane (Chen and Garwood, 1973a). Shown in Figure 5.4 is the product distribution from n-butane conversion over erionite at 21 atm, 1.5 LHSV over a temperature range of 400–510°C.

A commercial process converting n- and iso-butanes selectively to propane was developed by IFP (Fromager et al., 1977) using mordenite containing a non-noble metal hydrogenation function (Bernard et al., 1976). The range of operating conditions for this process is similar to that of Selectoforming.

Table 5.3 Selective Cracking for Propane and Selectoformate

Case:	1		2	
Yields, wt %	Charge	Products	Charge	Products
C_1, C_2, H_2	7.0	8.7	13.6	16.6
C_3	5.3	15.3	10.8	15.5
C_4	5.9	6.3	10.8	8.9
i-C_5	6.4	7.6	8.7	8.0
n-C_5	6.1	1.6	7.0	3.9
C_6 + n-paraffins	8.9	1.8	2.6	1.3
C_6 + iso's + aromatics	60.4	58.7	46.5	45.8
Octane numbers	Reformate	Selectoformate	Reformate	Selectoformate
C_5^+, RON + 0	86.1	93.3	96.5	99.5
C_5^+, RON + 3	96.3	100.8	102.6	104.7
C_5^+, DON + 3[a]	93.0	98.5	100.2	103.9
C_6^+, RON + 0	87.3	94.3	102.9	104.7
C_6^+, RON + 3	97.1	101.3	106.3	107.6

[a]Distribution octane number: a measure of front-end octane quality.
Source: Chen et al. (1968).

Table 5.4   Selectoforming a C_5-82°C Light Naphtha

28 atm, 1.6 LHSV

| | | Product at | |
Composition	Feed	395°C	425°C
Methane + ethane		1.5	5.0
Propane		21.9	24.7
Isobutane		0.4	1.3
n-Butane		6.0	7.7
Isopentane	22.2	22.1	21.7
n-Pentane	23.4	11.1	2.8
C_6^+	54.4	37.0	31.5
C_5^+ Octane, R + 0	67.4	79.6	84.0

Source: Heck (1987).

M-Forming

A new post-reforming process using ZSM-5 as the catalyst, known as the M-Forming process (Chen, 1973; Heinemann, 1977) was developed in the early 1970s at Mobil. The catalyst performs two major functions, to selectively crack paraffins and to alkylate benzene and toluene present in the reformate with the olefinic portion of the cracked products. The rate of alkylation favors benzene and toluene over the higher alkylaromatics in the reformate (Chen and Garwood, 1978b; Garwood and Chen, 1980). These lighter alkylated products provide the additional liquid yield and retain the high octane rating of benzene and toluene, while heavy aromatics outside the gasoline boiling range are not produced to any significant extent due to shape sepective constraints. The process improves liquid yield and reduces the concentration of benzene in the product.

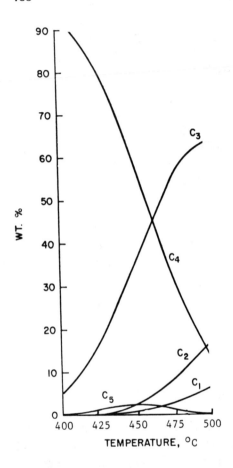

Figure 5.4   Product distribution from hydrocracking n-butane. (From Chen and Garwood, 1973a.)

The process may be operated with or without hydrogen circulation. However, in commercial practice as part of conventional reforming, it is convenient to integrate the M-Forming process into the reforming loop with continued hydrogen circulation.

As discussed in Chapter 3, unlike that for erionite, the rate of cracking of paraffins over ZSM-5 increases with the length of the molecules, and decreases with the bulkiness of the molecules (see Figure 4.9). For octane boosting, this is a most desirable feature, because the octane rating of paraffinic molecules increases in a similar fashion. Figure 5.5 shows the relationship between the relative cracking rate of C_5 to C_7 paraffins and their octane ratings.

To obtain yield improvement with M-Forming, it is important to properly interface the process with reforming. Generally speaking, this interfacing may be done on the basis of the octane-to-yield relationship. M-Forming is best interfaced at an octane severity beyond which reforming yield loss becomes excessive. This may depend on the composition and the boiling range of the naphtha, and the specific reforming process. On the other hand, M-Forming may indirectly influence the choice of the reforming process, e.g., a combination of ultra-low pressure reforming at moderate severities and M-Forming could achieve improved liquid yield and cycle life than current reforming processes that encounter catalyst stability problems at high reaction severities.

Figure 5.6 shows a typical yield/octane curve comparing regular reforming at 28 atm with an added M-Forming step (Bonacci and Patterson, 1981, Chen et al., 1987). In this case, the feed is a C_6-80°C mid-continent light naphtha. The "interface" between reforming and M-Forming was made at 85 clear octane (C_5^+). It is noted that reforming beyond this "interface" point gave a lower liquid yield than M-Forming.

Table 5.5 shows detailed compositional changes at two different M-Forming severities. Aromatic alkylation is indicated by the disappearance of benzene and toluene and the appearance of C_8^+ aromatics.

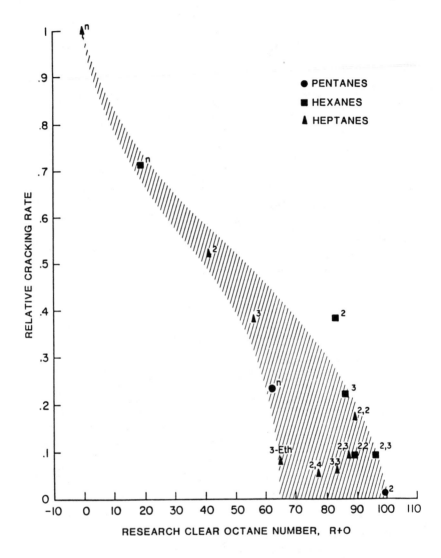

Figure 5.5   Octane rating vs. cracking rate of paraffins. (From Chen et al., 1979.)

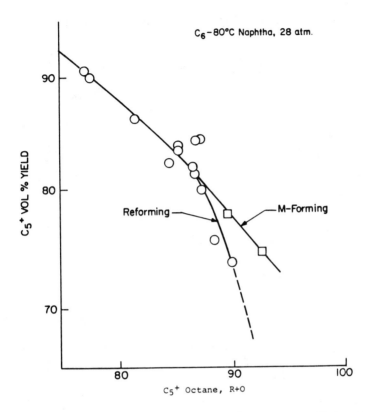

Figure 5.6   Reforming vs. M-Forming.   (From Chen et al., 1987.)

    For higher boiling naphthas, the "interface" is usually much
higher, above 90 R+0, in order to achieve yield advantage over
reforming.
    Shown in Figure 5.7 are the results obtained by interfacing
M-Forming with reforming a full range C_6 to 165°C Persian
Gulf naphtha at 76-, 86- and 91-octane severities. It is noted
that the octane of the M-formate increased at a nearly constant
rate of about one octane for every 1.4 vol % loss of C_5^+ yield
irrespective of the octane of the reformate. The yield loss by
reforming for this naphtha is less than that of M-Forming up to
about 94 R+0. Beyond 94 R+0, the yield loss by reforming

Table 5.5   Compositional Change at Different M-Forming Severities

$C_5{}^+$, R+0	C_6-80°C mid-continent naphtha, 28 atm				
	Feed 84.5	Product 89.6		Product 92.7	
Aromatics					
B	18.5	16.9	-1.6	16.3	-2.2
T	23.4	22.5	-0.9	21.7	-1.7
X	0.6	1.1	+0.5	1.5	+0.9
C_9	0.2	2.9	+2.7	3.7	+3.5
C_{10}	0	2.0	+2.0	3.3	+3.3
C_{11}	0	0.3	+0.3	0.4	+0.4
Total	42.7	45.8	+3.0	46.8	+4.1
Paraffins					
C_6	30.9	26.2	-4.7	23.4	-7.5
C_7	12.9	9.7	-3.2	8.7	-4.2
C_8	0.2	0	-0.2	0.3	+0.1
C_9	0	0	0	0	0
Total	44.0	35.9	-8.1	32.4	-11.6
Naphthenes	1.4	1.6	+0.2	1.1	-0.3
Pentanes	8.0	8.6	+0.6	9.0	+1.0
Butanes and ligher	3.9	8.1	+4.2	10.7	+6.8

Source: Chen et al. (1987).

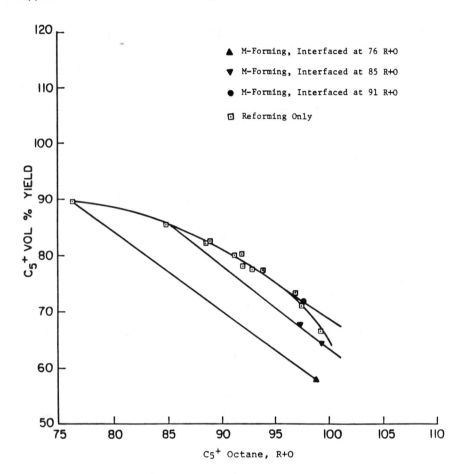

Figure 5.7   Reforming/M-forming C_6-165°C Persian Gulf naphtha. (From Chen et al., 1987.)

Table 5.6   Comparison of M-Forming and Reforming of a
C_6-165°C Persian Gulf Naphtha

	M-forming product	Reforming product	
C_5^+ R+O	96.8	96.7	
Aromatics, wt %			
benzene	5.5	5.5	0
toluene	15.7	16.3	−0.6
C_8	18.9	19.0	−0.1
C_9	11.0	17.5	−6.5
C_{10}	3.4	2.9	+0.5
C_{11}	1.3	0.1	+1.2
Subtotal	55.8	61.4	−5.6
isopentanes	7.9	6.9	+1.0
n-pentane	3.6	4.8	−1.2
isohexanes	13.9	13.5	+0.4
n-hexane	1.6	5.6	−4.0
isoheptanes	8.0	5.3	+2.7
n-heptane	0.1	1.5	−1.4
isooctanes	4.3	0.9	+3.4
n-octane	0	0.1	−0.1
isononanes	1.3	0	+1.3
n-nonane	0	0	0
isodecanes	0.1	0	+0.1
n-decane	0	0	0
naphthenes	3.4	0	+3.4
Subtotal	44.2	38.6	+5.6
Total	100	100	

Source: Chen et al. (1987).

becomes progressively greater than by M-Forming. Therefore, when the target octane exceeds 98 R+0, M-Forming clearly has a yield advantage over reforming alone.

Compared to conventional reformates at the same octane rating, the M-Forming products are low in aromatics and high in isoparaffins. Table 5.6 shows a typical example. The feedstock was a C_6-165°C Persian Gulf naphtha, the reforming was carried out at 18 atm over a Pt-Re bimetallic catalyst, and the product of M-Forming was made after reforming the naphtha to 84 clear octane and continued with M-Forming to 96.8 R+0. Interfacing at such a low octane severity did not achieve a yield advantage for M-Forming, however, the product composition clearly demonstrated the reduction in aromatics compared to the same naphtha reformed to the same octane.

II. CATALYTIC CRACKING OF GAS OILS

The addition of a small concentration of ZSM-5 to the fluid catalytic cracking reactor or the incorporation of ZSM-5 in the conventional catalyst boosts the octane rating of the gasoline from the catalytic cracking units by 1 to 3 numbers by selectively cracking the low octane paraffinic component to C_3 and C_4 olefins and paraffins (Anderson et al., 1984). The latter can be fed to an isobutane alkylation unit to produce additional high octane gasoline, raising the octane pool of the refinery (Yanik et al., 1985).

The process has been tested commercially in both TCC and FCC units. Results of a commercial test conducted in a 15,000-B/D TCC unit are shown in Table 5.7. In this test, the regular makeup catalyst (about 2 tons/day) was replaced by the new catalyst containing ZSM-5 for 93 days followed by the regular makeup catalyst. It was noted that even at 34 days after the addition of the ZSM-5 containing catalyst was stopped, the catalyst inventory (about 350 tons) in the unit continued to show octane-enhancing properties.

Similar results (Table 5.8) have been obtained commercially in FCC units (Yanik et al., 1985).

Table 5.7 Commercial Test of ZSM-5 Containing Cracking Catalysts in TCC

	Base catalyst	Catalyst containing ZSM-5	
Time on stream, days	0	72	106
Conversion, vol %	53.0	53.0	53.0
C_5^+ gasoline, vol %	42.3	40.9	41.0
Butenes, vol %	3.8	4.6	4.2
Propene, vol %	3.7	4.7	4.2
Light fuel oil, vol %	29.9	29.1	26.4
Research octane number R+0	86.0	90.2	91.2
Motor octane number M+0	77.4	79.2	79.5
Potential alkylate, vol %	12.7	15.7	14.2
Total gasoline, vol %	55.0	56.6	55.2

Source: Anderson et al. (1984).

Table 5.8 Commercial Test of ZSM-5 Containing Cracking Catalyst in FCC

	Base catalyst	Catalyst containing ZSM-5
Time on stream, days	0	37
Conversion, vol %	73.2	73.2
C_5^+ gasoline, vol %	53.9	51.6
Butenes, vol %	7.9	9.0
Propene, vol %	6.2	7.0
Light fuel oil, vol %	26.5	26.5
Coke, wt %	6.7	6.7
Research octane number R+0	87.3	89.0
Motor octane number M+0	78.1	78.7
Potential alkylate, vol %	24.9	28.3
Total gasoline, vol %	78.8	79.9

Source: Yanik et al. (1985).

III. DEWAXING OF DISTILLATE FUELS AND LUBE BASESTOCKS

Figure 5.8 shows the boiling ranges and associated average carbon numbers for current distillate fuels (Weisz and Katzer, 1984; Wise et al., 1986). With current boiling point specifications on diesel fuel in the U.S., use of pour point depressants and kerosene blending are required in the cold winter months to maintain acceptable fluidity. The cold flow property (pour point, freezing point, cloud point, and cold filter plugging point) of distillate fuels is largely determined by the concentration of normal and slightly branched paraffins in the oil. Figure 5.9 shows the melting point of pure paraffins as a function of carbon number and boiling point for the normal paraffins. As discussed in Chapter 4, the use of 8-membered-oxygen-ring small pore zeolites, such as erionite, in shape selective cracking appears to be limited to gasoline range molecules. Medium pore zeolites, such as ZSM-5, have been found to be effective in dewaxing a wide variety of feedstocks, ranging from jet fuels to resids, because they have just the pore size to discriminate these molecules from other bulkier molecules.

Synthetic ferrierite, a dual pore zeolite with interconnecting channels of elliptical 10- and 8-membered oxygen ring openings, and mordenite, a dual pore zeolite with interconnecting channels of puckered 12- and 8-membered oxygen ring openings, have also been found to be effective lube dewaxing catalysts with the incorporation of a noble metal hydrogenative function (Winquist, 1982; Bennett et al., 1975). Because its larger channels are unidirectional and the smaller channels are accessible only by the smaller molecules, they are not as effective as the medium pore zeolites for heavier feedstocks.

A. Mobil Distillate Dewaxing Process (MDDW)

By the selective removal of waxy molecules from the oil using the medium pore zeolites, the process improves the fluidity of

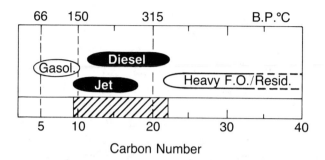

Figure 5.8   Carbon number and boiling-point range of premium petroleum products. (From Wise et al., 1986.)

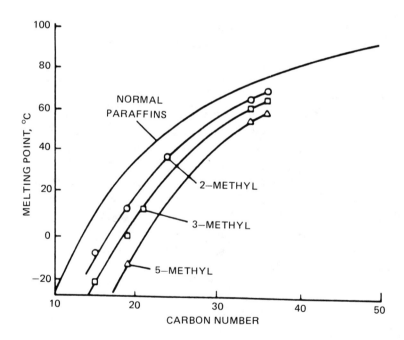

Figure 5.9   Melting points of normal and branched paraffins. (From Wise et al., 1986.)

the oil at low temperatures. It is the basis of the Mobil Distillate Dewaxing process, first announced in 1977 (Chen et al., 1977; Meisel et al., 1977). The process was commercialized in 1974 and is now in commercial operation world-wide, including the U.S., Canada, Italy, Germany, Indonesia, and China. The process is available for licensing from Mobil.

MDDW is a fixed bed process. Figure 5.10 shows a schematic flow diagram of an MDDW unit. The process is characterized by its long operating cycles (6 months to 1 year) between regenerations. Typical operating conditions for the MDDW process are 20 to 55 atm reactor pressure, 260–430°C reactor temperature, and 40–70 m^3/B of hydrogen circulation (Chen et al., 1977; Perry et al., 1978; Ireland et al., 1979; Graven and Green, 1980; Donnelly and Green, 1980). A typical product yield from an Arab Light Atmospheric Gas Oil (24.1 API gravity, and +15°C pour point) is shown in Table 5.9. The by-product gasoline, which represents two-thirds of the cracked product, has a high enough octane rating to be blended directly into the gasoline pool. Of course, distillate product yield decreases and gasoline yield increases as the wax content of the chargestock increases, and also as the pour point of the product decreases.

Feedstock to the MDDW process does not require hydro-treating, because the catalyst can selectively exclude most of the nitrogen and sulfur compounds, some of which are known as catalyst poisons. After going on stream, the reactor temperature required to produce target pour product initially increases fairly rapidly to a line-out temperature.

Historically, cold flow properties determine the end point of distillate fuels, but over the years, the maximum end point became an independent specification in many countries. Depending on the crude source, and the climate conditions of the markets served, many refineries must undercut the end point of No. 2 fuel oil and diesel fuel below specification boiling point limits in order to meet fluidity requirements. With MDDW, not only would undercutting be unnecessary, high economic incentive exists for converting a portion of the heavy distillate from the residual fuel pool into marketable distillate fuels. Thus, this

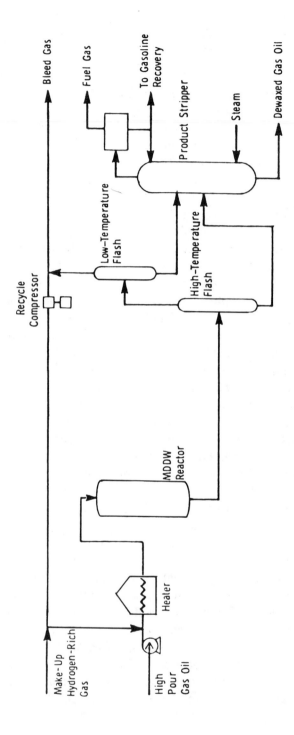

Figure 5.10 MDDW process schematic. (From Wise et al., 1986.)

Table 5.9   Pilot Plant MDDW Yield,
Arab Light Atmospheric Gas Oil
Charge

Distillate pour point, $-25°C$	
	Yield, wt %
Ethane and lighter	0
Propane	1.3
Butanes	3.3
Pentanes	2.9
C_6-165°C gasoline	5.9
165°C$^+$ distillate	86.4
Gasoline, R+0	90

Source: Donnelly and Green (1980).

new technology offers an opportunity to untie the historical
relationship between cold flow properties and end point of
distillate fuels and to increase the amount of high value distillates with acceptable pour points producible from a given crude,
if the end point of the product is relaxed. Extensive field tests
of extended-boiling-range dewaxed distillates conducted in
various European countries (Chen et al., 1977; Ireland et al.,
1979) showed similar combustion performance to conventional
diesel fuels and heating oils. So far, end point relaxation has
taken place in European countries, but not in the United States.
      By removing the wax from the high-pour-point distillate,
its pour point blending characteristic is changed from that of
virgin gas oils. It is well known that the pour point of virgin
distillates blends nonlinearly (Reid and Allen, 1951), i.e., the
pour point of a blend is higher than that of a calculated weighted
linear average value. When a high pour virgin distillate is dewaxed
and then blended with a low-pour-point virgin distillate, the
blend has an expected lower pour point than that a calculated

weighted linear average. This is illustrated in Figure 5.11.
Advantage can be taken of the favorable blending characteristic
of catalytically dewaxed oils by fractionating a gas oil, dewaxing
only the higher boiling fraction and then blending these two
streams.

In addition to processing high-pour-point virgin gas oils, the
MDDW process may be integrated with catalytic cracking process
especially for processing gas oils from very waxy crudes by
either dewaxing of FCC cycle oils or diverting the light portion
of the catalytic cracker feed to the MDDW, unloading the
catalytic cracker for processing heavier feed (Perry et al.,
1978; Pavlica et al., 1980; Donnelly and Green, 1980;
Graven and Green, 1980). This is particularly desirable

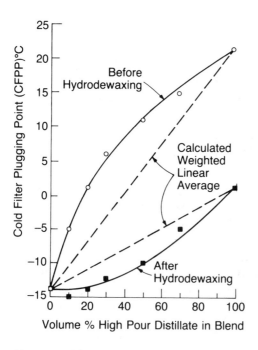

Figure 5.11   Synergistic pour-point blending of hydrodewaxed
high-pour distillate. (From Chen et al., 1978.)

when the market for distillate fuel exceeds that of the gasoline market.

One of such processing schemes is shown schematically in Figure 5.12. In this scheme, the 350-415°C fraction of the FCC feedstock is diverted to the MDDW unit. This allows the FCC unit to operate at higher recycle rates and higher distillate yields. The gasoline-to-distillate ratio of a typical FCC unit operating in the distillate mode may be as low as 1.3. With the addition of a MDDW unit, laboratory data shows that it is possible to reduce the gasoline-to-distillate ratio produced to below 0.7 (Table 5.10).

Of course in order to realize the full benefit of this scheme, it is necessary to relax the end point of the distillate fuel, but not sacrificing the low-temperature-fluidity requirements.

When processing high sulfur, high nitrogen distillates (Gorring and Shipman, 1975), it is preferable to dewax the gas oil first and cascade the entire effluent from the dewaxing reactor to a conventional hydrotreating reactor to avoid the adverse effect of ammonia and hydrogen sulfide on the dewaxing catalyst and the need to have a two stage process with interstage separation. The cascade process saves a significant amount of process energy (Shen, 1983) over the two-stage process.

The efficiency of the freezing point/pour point lowering over ZSM-5 increases with conversion (Chen et al., 1979). This nonlinear relationship between pour point and conversion is illustrated in Figure 5.13. Complete extraction of the 12% n-paraffins in the charge by shape selective sorption in 5 Å zeolite lowered the pour point to only −18°C. Since more than 12% of the charge is being converted over ZSM-5 to reduce its pour point to −18°C and lower, other high-pour-point molecules, primarily isoparaffins and a lesser amount of selected 1-ring aromatics and naphthenes are converted.

At −25°C pour, only a trace of n-paraffins remained (Table 5.11). Once the n-paraffins are gone, the pour point dropped dramatically with only a small additional conversion to C_{17}^- product. This is the subtle change that accounts for the non-linearity of pour point with conversion.

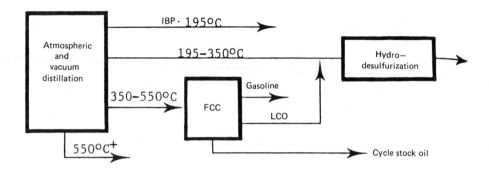

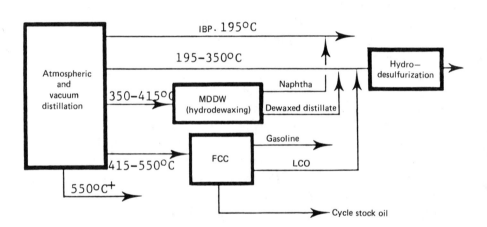

Figure 5.12   Distillate fuel manufacturing schemes. (From Perry et al., 1978.)

Table 5.10   Impact of MDDW on Distillate-to-Gasoline Ratio

Feed boiling range, °C:	FCC	FCC + MDDW
FCC	350–500	410–550
MDDW	—	350–410
Gasoline/distillate ratio	1.3	0.7
Distillate fuel composition, vol % light virgin gas oil	81	59
FCC light cycle oil	19	23
MDDW product	—	18

Source: Perry et al. (1979).

Selected alkylbenzenes and alkylnaphthenes are converted to lower boiling range and appear as a high octane naphtha product (Table 5.12).

Figure 5.14 shows a more detailed pattern of conversion of various classes of paraffins and alkylbenzenes as a function of

Table 5.11   Sensitivity of Pour Point to Paraffin Content

C_{17}–C_{25} heavy gas oil, 35 atm, 370°C					
			Charge		
Pour point, °C	+10	0	−25	−40	−60
Paraffin content, wt %	36.8	30.8	25.2	23.6	21.8
n-Paraffins detectable by M.S.	Yes	Yes	Trace	No	No
Incremental pour point lowering percent paraffin loss	—	3.3	7.1	15.6	25.0

Source: Chen et al. (1979).

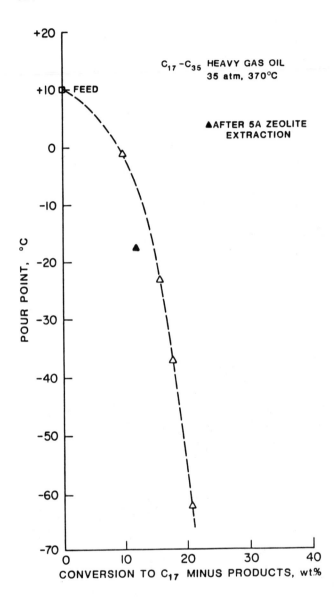

Figure 5.13   Pour point vs. conversion.   (From Chen et al., 1979).

Table 5.12   Compositional Change Before and After Dewaxing

	Feed: C_{17}-C_{25} heavy gas oil		
	Feed +10°C pour	Product, 69 wt % -60°C pour	Wt % converted
Aromatics, wt %	26.9	36.7	5.9
alkylbenzenes	6.1	7.1	20
tetralins	5.8	8.2	2
naphthalenes	3.5	4.9	3
acenaphthenes	4.1	5.8	2
high polynuclears	7.4	10.7	Nil
Naphthenes	35.8	45.6	11.9
1-ring	17.4	20.4	19
2-ring	9.7	13.1	7
3-ring	4.6	6.2	6
4-5 ring	4.1	6.0	Nil
Paraffins	37.3	17.6	67.4
C_{17}	4.1	1.4	76
C_{18}	3.7	1.5	72
C_{19}	5.5	0.6	92
C_{20}	5.2	3.2	58
C_{21}	5.1	4.1	45
C_{22}	4.4	1.6	75
C_{23}	3.0	1.8	59
C_{24}	2.7	1.0	74
C_{25}	1.6	1.2	48
C_{26}^{+}	2.0	1.2	59

Source: Chen et al. (1979).

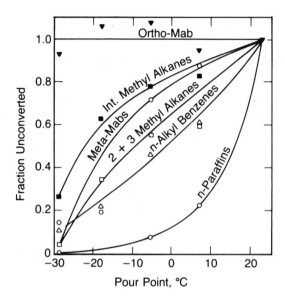

Figure 5.14   Conversion of alkane and alkylbenzene molecular classes during dewaxing of chargestock A over ZSM-5. Mab = methylalkylbenzene. (From Bendoraitis et al., 1986.)

the pour point of the dewaxed product. Examination of the trends for paraffin conversion shows that all n-paraffins in the gas oil (boiling range 310–455°C) are readily converted, followed by monomethyl paraffins. Paraffins containing quaternary carbons or polymethyl substituted are not converted. The conversion pattern of the methyl alkylbenzenes indicates a sharp cut-off between the meta- and ortho-isomers which differ less than 0.5 Å in their critical molecular dimensions.

B. Jet Fuel Dewaxing

Jet fuels are a special type of distillate fuels high in flash point (>170°C) to reduce fire hazard during refueling; high in heating value for long flying range; and low in freezing point for high altitude. Traditionally, they were produced from petroleum by distillation and purification with little boiling range conversion or catalytic upgrading. Thus, by necessity, they are crude-dependent and are limited to very narrow boiling ranges. For example, the commercial jet fuel, Jet A, has a boiling range of 170°C to 270°C and a freezing point of -40°C, and the premium military jet fuel, JP-7, specifies an even narrower boiling range because it also requires a minimum heating value of 18,750 BTU/lb, which limits its aromatics content to 5 vol % maximum. To meet these requirements, its hydrocarbon composition is confined to a relatively narrow region rich in paraffins as shown in Figure 5.15.

By combining catalytic dewaxing with product hydrogenation, it is now possible to produce these jet fuels from a wide range of feedstocks.

Using the small pore zeolite, Ni-erionite, described previously in the Selectoforming process, jet fuel specification product has been made from a -40°C freezing point paraffinic feed. However, the dewaxing reaction was found to be dependent on hydrogen pressure (Chen and Garwood, 1978a). Freezing point lowering was effected at >69 atm of hydrogen pressure (Table 5.13). The same feed processing over ZSM-5 requires only a moderate pressure of 34 atm.

As shown in Figure 5.16, the freezing point of the jet fuel product from ZSM-5 processing, as in the case of pour point lowering of distillate fuels, decreased nonlinearly with the extent of boiling range conversion. At low conversions, the freezing point of the product decreased only slightly. But as the freezing point was lowered to below -40°C, each additional percent increase in conversion led to a drop of more than 2.3°C in the freezing point.

With catalytic dewaxing, it is also possible to convert a portion of the heavier distillate to jet fuels (Chen and Garwood, 1986). For example, the supply of Jet A could be increased 25% or more by the inclusion of 260-290°C boiling range kerosene. Shown in Table 5.14 are the results of dewaxing a 177-260°C and a 260-290°C kerosene fractions. By blending the

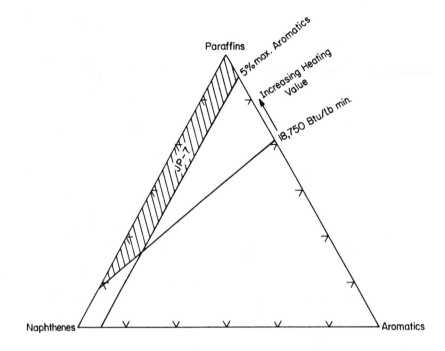

Figure 5.15   Chemical composition of JP-7 jet fuel (vol %).
(From Chen and Garwood, 1987.)

high-boiling-range product with the standard fuel, potentially all
of the 260–290°C fraction available from most crudes could be
upgraded to jet fuel and meet all the quality requirements.

Premium military jet fuel, JP-7, requires a ≤45°C freezing
point and a minimum heating value of 18,750 BTU/lb, which
can be met only with paraffinic stocks. Waxy crudes such as
Libyan crude, being highly paraffinic, are potential sources for
the manufacture of high energy jet fuels except for their high
freezing points. With catalytic dewaxing over ZSM-5 combined
with product hydrogenation, JP-7 specification fuel can be
produced. A comparison of JP-7 specification with the product
from the combined catalyst system is given in Table 5.15.

Table 5.13   Effect of Pressure on Freeze Point of Jet Fuel

Ni/H-erionite, 400°C, 14 LHSV, 27/1 H_2/hydrocarbon				
Pressure, atm	14.6	35	69	137
Conversion, wt %				
n-paraffins	15.6	16.2	58.2	73.8
non-normals	12.1	18.2	21.8	22.7
Jet fuel yield, wt %	87.0	82.3	69.1	64.6
Freeze point, °C	−38	−39	−47	−53
Δ Freeze point	+2	+1	−7	−13
Freeze point depression efficiency °C/% conversion			0.23	0.37

Source: Chen and Garwood (1987); and Chen and Garwood (1978a).

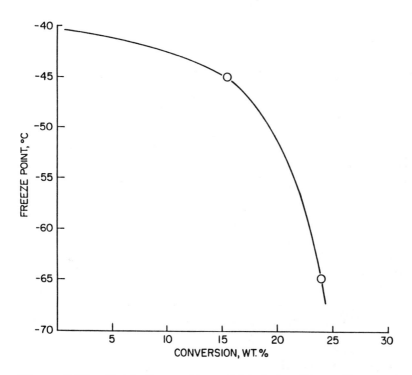

Figure 5.16   Catalytic cracking of jet fuel. (From Chen and Garwood, 1987.)

Table 5.14   Catalytic Dewaxing of 177–260°C and 260–290°C
Kerosene Fractions

	Zn/H-ZSM-5, 34 atm, 24 LHSV	
Charge stock:	177–260°C (Stock B)	260–290°C (Stock C)
H$_2$/HC mole ratio	5	7
Temperature, °C	340	370
Conversion, wt %	13.4	22.0
Jet fuel product		
freezing point, °C	–65	–43
KV at –35°C, cs	8.52	20.70

Source: Chen and Garwood (1987).

Table 5.15   Catalytic Dewaxing Followed by Hydrogenation

	Libyan 177–260°C kerosene			
	Charge stock E	Dewaxed[a] product	Then[b] hydrogenated	JP-7 specification
Gravity, °API	48.1	47.2	47.5	44–50
Freezing point, °C	–33	–53	–54	–45 max
Aromatics, vol %	9.1	11.9	4.0	<5
Heating value, BTU/lb	18,710	18,655	18,835	18,750 min

[a]Zn/H-ZSM-5, 343°C, 34 atm, 15 H$_2$/HC mole ratio, 24 LHSV. 78.5 wt % yield
of dewaxed product.
[b]Ni/Kieselguhr, 34 atm, 315°C.
Source: Chen and Garwood (1987).

C. Mobil Lube Dewaxing Process (MLDW)

Mobil's lube dewaxing process differs from the fuels dewaxing process in that it utilizes a two-reactor system. A schematic process flow diagram is shown in Figure 5.17. The first reactor contains the ZSM-5-based catalyst. The effluent from the first reactor is cascaded to the second reactor without product separation. The second reactor contains a hydrofinishing catalyst to assure that the final lube product will meet all quality and engine performance specifications (Graven and Green, 1980; Smith et al., 1980).

For solvent-refined raffinates, the catalyst may age at one or more degrees centigrade per day making the process cyclic, with cycle lengths lasting several weeks to several months. For hydrotreated feeds, the process is operated in the steady state mode with cycle length exceeding a year.

The process can be used to produce a full slate of lube basestocks, ranging from the conventional lubes, and hydro-cracked lubes to very-low-pour-point products required for transformer and refrigeration services. It offers many advantages over the solvent dewaxing technology, including lower capital investment and lower operating cost. The feed to MLDW can be properly stripped raffinates (Gillespie et al., 1979; Gillespie et al., 1980; Gillespie et al., 1984), from any of the solvent extraction processes (furfural, phenol, N-methyl-2-pyrrolidone) commonly practiced in lube manufacture to remove aromatics, or raw distillates, soft waxes, and deasphalted resids.

The combination of commercial lube hydrotreating process and MLDW offers for the first time the potential of developing an all-catalytic route to lubes. Several companies that have such hydrotreating facilities have added catalytic dewaxing for their lighter stocks (Garwood and Silk, 1981; Wise et al., 1986).

If economic incentives are great enough to justify wax production, a combination of solvent dewaxing with MLDW can be attractive (Chen and Garwood, 1973b; Smith et al., 1980). Saleable wax can be recovered by partial solvent dewaxing at an intermediate temperature above the freezing point of water.

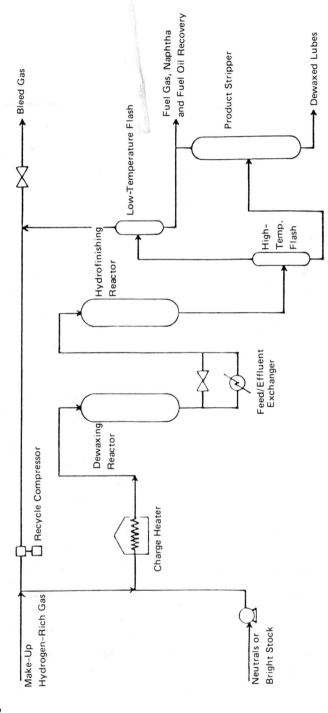

Figure 5.17 Simplified MLDW process flow diagram. (From Graven and Green, 1980.)

Partial solvent dewaxing is much less costly than complete solvent dewaxing because of higher filter rates, easier wax deoiling, and reduced refrigeration requirements. Foots oil, a waxy by-product of solvent dewaxing, can also be suitable stock for catalytic dewaxing (Garwood and Wise, 1976).

Typical lube yields from the MLDW process are shown in Table 5.16. The yields are based on processing an Arab Light crude. The by-product yield is similar to that of the distillate dewaxing process, i.e., one-third LPG and two-thirds naphtha.

Physical properties of catalytically dewaxed basestocks are similar to that of the solvent dewaxed stocks with the exception of viscosity and viscosity index. VI is traditionally the parameter used to control the severity of the aromatics extraction step that precedes dewaxing. With the same extraction severity, the viscosity of the MLDW product is higher and the viscosity index lower than that of the SDW product. These differences do not, however, adversely affect end-use performance of the finished oils. The difference decreases with increasing basestock viscosity and is negligible with bright stocks.

Extensive product quality tests have been conducted on various finished products, including engine oils, industrial turbine and circulating oils, gear oil, refrigerator and transformer oils, made by blending catalytic dewaxed basestocks and compared with those made with solvent dewaxed basestocks. In all tests, products containing the catalytic dewaxed oil blends show equal or better performance than their solvent-dewaxed counterparts. Table 5.17 shows the test results for engine oils.

The MLDW process was tested on a large scale in Mobil's Gravenchon Refinery in France in 1978. The first grass-root commercial MLDW unit went on stream in 1981. The process is also available for license from Mobil and is currently in commercial operation in several countries, including the U.S., Australia and Japan.

An all-catalytic, hydrocracking-dewaxing process has recently been commercialized by Chevron for producing high-viscosity-index lubes (Farrell and Zakarian, 1986; Zakarian et al., 1987).

Table 5.16   Commercial MLDW Lube Yields

Feed: Arab Light raffinates	Yield, wt %
100 SUS light neutral	74.5
300 SUS heavy neutral	79.8
700 SUS heavy neutral	82.2
Bright stock	91.2

D. Other Catalytic Dewaxing Processes

The catalytic dewaxing process developed by the British
Petroleum Company (Donaldson and Pout, 1972; Bennett et al.,
1975; Hargrove et al., 1978) is a catalytic hydrocracking process,
which employs a bifunctional platinum/H-mordenite catalyst

Table 5.17   Engine Oil Quality Tests

Marine diesel engine oil:	SDW	MLDW	
Caterpillar 1-G			
total weighted demerits	70	47	100 max
total groove fill, %	18	18	25 max
Passenger car engine oil:			
Ford MS sequence VC			
sludge	8.5	9.5	8.5 min
varnish	8.2	8.8	8.0 min
piston skirt varnish	8.0	8.3	7.9 min
cold cranking simulator cp at $-18°C$	2400	2400	2400 max

Source: Smith et al. (1980).

(Burbidge et al., 1971). The fixed bed process is operated under hydrogen pressure to saturate the cracked by-products consisting largely of propane, butanes and pentanes. High hydrogen pressure also minimizes coke formation and maintains stable catalyst activity. This is particularly important with mordenite, which contains unidirectional channels prone to coking when used as an acidic catalyst. The catalyst has very little activity for desulfurization or denitrogenation.

The process is applicable to the production of low-pour-point distillate fuels and very-low-pour-point low viscosity oils from naphthenic and partially dewaxed paraffinic distillates, but is not suitable for the production of high-viscosity-index, high viscosity lubricating oils. Dewaxed products with a pour point as low as $-50°C$, suitable as transformer and refrigerator oils, and low-pour-point, extended-boiling-range diesel fuel and No. 2 heating oils have been produced.

A similar catalytic dewaxing process employing a bifunctional Pd/H-ferrierite catalyst was developed by Shell (Winquist, 1982). The process can be combined with a solvent dewaxing process using methylisopropyl ketone as the solvent, which produces an intermediate-pour-point stock for the catalytic dewaxing process (Stem, 1986).

IV. DISTILLATE FROM LIGHT OLEFINS (MOGD)

The oligomerization, transmutation/disproportionation and aromatization of C_2 to C_{10} olefins, discussed in Chapter 4, form the basis of Mobil's Olefin to Gasoline, Distillate Process (MOGD) (Tabak, 1984; Tabak et al., 1984; 1986). In the distillate mode, the process is operated under elevated pressures and low temperatures to produce iso-olefins; in the gasoline mode, the process is operated at higher reaction temperatures to increase aromatics production. Figure 5.18 shows that as the pressure is increased to above 23 atm, propene at 200°C is converted to $165°C^+$ olefins at better than 70% yield (Garwood, 1983). After hydrogenation, the products are premium quality low-pour-point jet

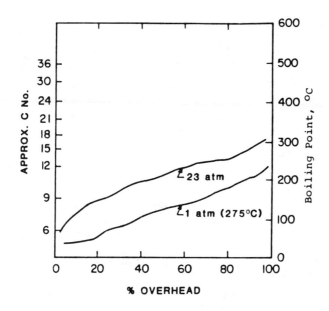

Figure 5.18 Effect of pressure on propene conversion, 0.4 WHSV, 200°C. (From Garwood, 1983.)

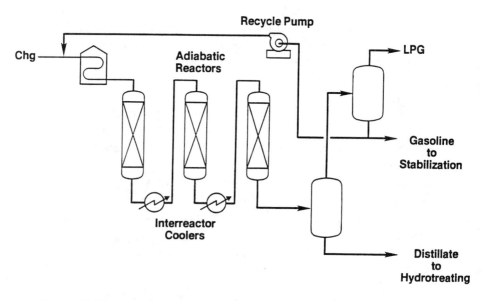

Figure 5.19 MOGD process flow max distillate mode. (From Tabak et al., 1986a.)

196

Table 5.18   Yield and Quality of the MOGD Products

	MOGD product	Industry standard
Diesel fuel		
Specific gravity, at 15°C	.79	.84-.88
Flash point, °C	60	52
Pour point, °C	≤50	-7
Cetane number	50	45
Sulfur, wt %	<.002	0.5 max
Viscosity, cs at 40°C	2.5	1.9-4.1
Boiling range, °C		
10%	203	249 max
50%	264	000 max
90%	324	338 max
Jet fuel		
Specific gravity	.79	.78-.8
Freezing point, °C	≤60	-40
Sulfur, wt %	<.002	0.3 max
Smoke point, mm	28	18 min
JETOT, °C	343	260
Hydrocarbon type,		
FIA aromatics	4	25 max
olefins	1	5 max
paraffins	95	—

Source: Tabak et al. (1986b).

fuel and distillate fuels. The selectivity of the catalyst plays a major role in producing these unique products. A schematic process flow diagram is shown in Figure 5.19. The design uses three reactors with interstage cooling and a condensed liquid recycle. While liquid recycle provides a heat sink for the control of reaction exotherm, it also affects product quality by allowing the recycled material to be further reacted. In this design, both maximum gasoline and maximum diesel modes can be run by adjusting the reactor temperature and the boiling range of the recycle stream. Heavier recycle streams favor the production of distillates.

Table 5.18 presents the yield and quality of the MOGD products. As diesel, the paraffinic MOGD product is low in density but is an exceptionally good blending stock due to its low pour point and negligible sulfur content. Because of its paraffinic nature, the MOGD product makes excellent jet fuels, meeting or exceeding all commercial and military specifications. In particular, because of its low aromatic content, it is a superbly stable fuel as indicated by the JFTOT test.

With the use of a C_3/C_4 mixed olefin stream from the refinery, the MOGD process was tested commercially in Mobil's Paulsboro Refinery in 1982; it is now offered for license by Mobil.

REFERENCES

Anderson, C. D., F. G. Dwyer, G. Koch, and P. Niiranen, "A New Cracking Catalyst for Higher Octanes Using ZSM-5," paper presented at the 9th Iberoamerican Symp. Catal., Lisbon, Port., July 1984.

ASTM (Am. Soc. Testing Materials), "ASTM Manual for Rating Motor, Diesel and Aviation Fuels," ASTM, Philadelphia (1973).

Banta, F., H. R. Ireland, T. R. Stein, and R. C. Wilson, U.S. Pat. 4,247,388, Jan. 27, 1981.

Bendoraitis, J. G., A. W. Chester, F. G. Dwyer, and W. E. Garwood, "Pore Size and Shape Effects in Zeolite Catalysis," paper presented at the 7th Intn. Zeol. Conf., Tokyo, Aug. 17–22, 1986.

Bennett, R. N., G. J. Elkes, and G. J. Wanless, Oil Gas J. *3*(1), 69 (1975).

Benson, J., Chemtech *6*, 16 (1976).

Bernard, J. R., J. Bousquet, and M. Grand, German Pat. 2,626,840, Dec. 23, 1976.

Bonacci, J. C. and J. R. Patterson, U.S. Pat. 4,292,167, Sept. 29, 1981.

Burd, S. D. and J. Majiuk, Oil Gas J. *70*(27), 52 (1972).

Burbidge, B. W., I. M. Keen, and M. K. Eyles, Advan. Chem. Ser. *102*, 400 (1971).

Chen, N. Y. and W. E. Garwood, U.S. Pat. 3,379,640, Apr. 23, 1968.

Chen, N. Y., J. Maziuk, A. B. Schwartz, and P. B. Weisz, Oil Gas J. *66*(47), 154 (1968).

Chen, N. Y. and E. J. Rosinski, U.S. Pat. 3,630,966, Dec. 28, 1971.

Chen, N. Y., U.S. Pat. 3,729,409, Apr. 24, 1973.

Chen, N. Y. and W. E. Garwood, Advan. Chem. Ser. *121*, 575 (1973a).

Chen, N. Y. and W. E. Garwood, U.S. Pat. 3,755,138, Aug. 28, 1973b.

Chen, N. Y., R. L. Gorring, H. R. Ireland, and T. R. Stein, Oil Gas J. *75*(23), 165 (1977).

Chen, N. Y. and W. E. Garwood, Ind. Eng. Chem. Process Des. Dev. *17*, 513 (1978a).

Chen, N. Y. and W. E. Garwood, J. Catal. *52*, 453 (1978b).

Chen, N. Y., B. M. Gillespie, H. R. Ireland, and T. R. Stein, U.S. Pat. 4,067,797, Jan. 10, 1978.

Chen, N. Y., W. E. Garwood, W. O. Haag, and A. B. Schwartz, "Shape Selective Hydrocarbon Conversion Over Synthetic Zeolite ZSM-5," paper presented at Symp. Advan. Catal. Chem. I, Snowbird, Utah, Oct. 3–5, 1979.

Chen, N. Y. and W. E. Garwood, Ind. Eng. Chem. Prod. Res. Devel. *25*, 641 (1986).

Chen, N. Y., W. E. Garwood, and R. H. Heck, Ind. Eng. Chem. Process Des. Dev. *26*, 706 (1987).

Donnelly, S. P. and J. R. Green, "Recent MDDW Process Improvements," paper presented at the Symp. Jap. Petrol. Inst., Tokyo, Oct. 2–3, 1980.

Donaldson, K. and C. R. Pout, "The Application of a Catalytic Dewaxing Process to the Production of Lubricating Oil Basestocks," paper presented at the 164th Am. Chem. Soc. Nat. Mtg., New York, Sept. 1972.

Farrell, T. R. and J. A. Zakarian, Oil Gas J. *84*(20), 47 (1986).

Fromager, N. H., J. R. Bernard, and M. Grand, Chem. Eng. Prog. *73*(11), 89 (1977).

Garwood, W. E. and J. J. Wise, U.S. Pat. 3,960,705, June 1, 1976.

Garwood, W. E. and N. Y. Chen, "Octane Boosting Potential of Catalytic Processing of Reformate Over Shape Selective Zeolite," paper presented at the Am. Chem. Soc. Mtg., Houston, Mar. 23–28, 1980; Div. Petrol. Chem. Preprint *25*(1), 84 (1980).

Garwood, W. E. and M. R. Silk, U.S. Pat. 4,283,271, Aug. 11, 1981; U.S. Pat. 4,283,272, Aug. 11, 1981.

Garwood, W. E., Am. Chem. Soc. Symp. Ser. *218*, 383 (1983).

Gorring, R. L. and G. F. Shipman, U.S. Pat. 3,894,938, July 15, 1975.

Gillespie, B. M., M. S. Sarli, and K. W. Smith, U.S. Pat. 4,147,148, Jan. 30, 1979; U.S. Pat. 4,181,598, Jan. 1, 1980; U.S. Pat. 4,437,975, Mar. 20, 1984.

Graven, R. G. and J. R. Green, "Hydrodewaxing of Fuels and Lubricants Using ZSM-5 Type Catalysts," paper presented at the 1980 Congr. Australia Inst. Petrol., Sydney, Sept. 15–17, 1980.

Hargrove, J. D., G. J. Elkes, and A. H. Richardson, "BP Cat. Dewaxing—Experience in Commercial Operation," paper presented at the NPRA Nat. Fuels and Lubricants Mtg., Houston, Nov. 9–10, 1978.

Heck, R. H., private communication, 1987.

Heinemann, H., Catal. Rev.-Sci. Eng. *15*, 53 (1977).

Ireland, H. R., C. Redini, A. S. Raff, and L. Fava, Hydrocarbon Process *58*(5), 119 (1979).

Meisel, S. L., J. P. McCullough, C. H. Lechthaler, and P. B. Weisz, "Recent Advances in the Production of Fuels and Chemicals Over Zeolite Catalysts," paper presented at the

Leo Friend Symp., 174th Am. Chem. Soc. Mtg., Chicago, Aug. 30, 1977.

Pavlica, R. T., T. R. Stein, and C. W. Streed, U.S. Pat. 4,192,734, Mar. 11, 1980.

Perry, R. H. Jr., F. E. Davis, Jr., and R. B. Smith, Oil Gas J. 76(21), 78 (1978).

Perry, R. J. Jr., C. Redini, A. S. Raff, and L. Fava, Hydrocarbon Process 58(5), 119 (1979).

Plank, C. J., E. J. Rosinski, and E. N. Givens, U.S. Pat. 4,141,859, Feb. 27, 1979.

Plank, C. J., E. J. Rosinski, and E. N. Givens, U.S. Pat. 4,276,151, June 30, 1981.

Ramage, M. P., K. R. Graziani, P. H. Schipper, and F. J. Krambeck, Advan. Chem. Eng. 13, 193 (1987).

Reid, E. B. and H. I. Allen, Petrol. Refiner 30(5), 93 (1951).

Roselius, R. R., K. R. Gibson, R. M. Ormiston, J. Maziuk, and F. A. Smith, "Rheniforming and SSC, New Concepts and Capabilities," paper presented at the NPRA Annual Mtg., San Antonio, Apr. 1–3, 1973.

Rosinski, E. J. and A. B. Schwartz, U.S. Pat. 4,309,280, Jan. 5, 1982.

Shen, R. C., U.S. Pat. 4,400,265, Aug. 23, 1983.

Smith, K. W., W. C. Starr, and N. Y. Chen, Oil Gas J. 78(21), 75 (1980).

Stem, S. C., U.S. Pat. 4,622,130, Nov. 11, 1986.

Tabak, S. A., "Production of Synthetic Diesel Fuel from Light Olefins," paper presented at the AIChE Nat. Mtg., Philadelphia, Aug. 1984.

Tabak, S. A., F. J. Krambeck, and W. E. Garwood, "Conversion of Propylene and Butylene Over ZSM-5 Catalyst," paper presented at the AIChE Mtg., San Francisco, Nov. 25–30, 1984; AIChE J. 32, 1526 (1986a).

Tabak, S. A., A. A. Avidan, and F. J. Krambeck, "Production of Synthetic Gasoline and Diesel Fuels from Non-Petroleum Sources," paper presented at ACS Nat. Mtg., New York, Apr. 13–15, 1986b.

Weisz, P. B. and J. R. Katzer, "Fuels from Carbonaceous Resources—Constraints and Opportunities," paper presented at CHEMRAWN III Conf., The Hague, June 25–29, 1984.

Winquist, B. H. C., U.S. Pat. 4,343,692, Aug. 19, 1982.

Wise, J. J., J. R. Katzer, and N. Y. Chen, "Catalytic Dewaxing in Petroleum Processing," paper presented at the Am. Chem. Soc. 173rd Annual Mtg., New York, Apr. 14–15, 1986.

Yanik, S. J., E. J. Demmel, A. P. Humphries, and R. J. Campagna, Oil Gas J. 83(19), 108 (1985).

Zakarian, J. A., R. J. Robson, and T. R. Farrell, Energy Prog. 7(1), 59 (1987).

6
Applications in Aromatics Processing

I. BTX SYNTHESIS

A. M2-Forming

As discussed in Chapter 4, ZSM-5 converts light paraffins, olefins, and naphthenes to aromatics and light gases. A generic name, M2-Forming has been coined to describe this new aromatization process (Chen and Yan, 1986). As shown by the data in Table 6.1, at 538 to 575°C and 1 WHSV, the aromatics yield from n-pentane and n-hexane is more than 30% by weight; similar yield is obtained from the more reactive propene at 68 LHSV; a virgin naphtha which contains 47% paraffins, 41% naphthenes, and 12% aromatics gives 44% aromatics or a net gain of 32%; and a light FCC gasoline that contains 41% of olefins and 10% aromatics gives 54% aromatics or a net gain of more than 44% aromatics.

In fact, all nonaromatic molecules are aromatizable. Only methane and ethane are not aromatized and their concentrations are continuously increased as the reaction severity is increased.

For paraffinic feed, the olefinic intermediates are mainly derived from the catalytic cracking step. With feedstocks containing olefins, diolefins, naphthenes and hydroaromatic compounds, and at relatively low temepratures direct aromatization via hydrogen transfer reactions may provide an alternate route to aromatics and saturates. At higher temperatures, however, catalytic cracking again dominates the first reaction step.

M2-Forming uniquely produces aromatic concentrates. Refinery streams rich in unsaturated hydrocarbons such as the by-product streams from the naphtha steam cracking process, known as pyrolysis gasoline or Dripolene, unsaturated gases

Table 6.1   Aromatization of Light Hydrocarbons

| | | Catalyst: ZSM-5, pressure: atmospheric | | | |
| | | Reaction | | Atmospheric yield, % | |
Feed	C%	temp, °C	WHSV	observed	extinction
n-Pentane	83.33	575	1	31	45
n-Hexane	83.72	538	1	32	45
Propene	85.71	538	68	33	58

Source: Chen and Yan (1986).

from the catalytic cracking process, and LPG are potential feed-stocks for this process. Because of the high concentration of aromatics produced, the product stream may be separated directly for BTX production without a costly additional solvent extraction step. Thus, the M2-Forming process could be an attractive alternative to the conventional hydrotreating/extraction process for upgrading pyrolysis gasoline to BTX (Chen and Yan, 1986).

Depending on the reactivity of the feedstock, the operating temperature varies significantly, ranging from less than 370°C for olefins and 538°C for propane.

Compared with other zeolites, the HZSM-5 catalyst is unique in its ability to catalyze the aromatization reaction with sustained activity over a period of several days. Its stability is attributed to the shape and size of the pore openings and the tortuosity of the channels which inhibit the formation of coke precursors (Chen and Garwood, 1978; Walsh and Rollmann, 1979). Nevertheless, because of eventual catalyst deactivation, periodic regenerations are necessary, thus the application of M2-Forming to aromatics production would be in the form of a cyclic process.

One of such cyclic aromatization processes, the Cyclar[TM] process, was recently announced by UOP and British Petroleum in 1984. A recent report (Oil Gas J., 1987) indicated that a world-first commercial scale unit using the Cyclar^{TM} process to convert LPG to aromatics, used as high octane gasoline blending components, will be built at the British Petroleum Grangemouth Refinery in Scotland. The process employs UOP's CCR reforming technology (continuous catalyst regeneration) to convert LPG to aromatics (Johnson and Hilder, 1984). Similar to naphtha reforming, the reaction is carried out in vertically stacked multi-stage adiabatic radial flow reactors with interstage heating. Figure 6.1 shows a schematic process flow diagram. In this process, the catalyst flow by gravity from one reactor to the next. After exiting the last reactor, the spent catalyst is transferred by lift gas to the top of the regenerator, where it is separated from the lift gas and flows by gravity again through the regenerator to a hopper. Lift gas is used to transfer the catalyst back to the top of the first reactor.

Kitagawa et al. (1986) reported a yield of 71% aromatics from propane through the addition of gallium to the ZSM-5 catalyst. Others have also reported improved yields of aromatics from propane by adding a metal function to ZSM-5 (Davies and Kolombos, 1979; Anderson et al., 1985; Mole and Anderson, 1985). Some of the phosphate-based molecular sieves also have been reported to have aromatization activity for converting olefins and diolefins to BTX aromatics (Garska and Lok, 1985).

B. Xylene Isomerization (MVPI)

The Mobil Vapor Phase Xylene Isomerization (MVPI) process is designed to isomerize the xylenes in a C_8 aromatics stream obtained from either reformates or pyrolysis gasolines. A schematic diagram of a xylene isomerization process is shown in Figure 6.2. The C_8 aromatics stream is fractionated in an ethylbenzene tower to remove ethylbenzene from the feed. Because the ethylbenzene separation step is energy-intensive, in recent years it has often been omitted. Thus the C_8 aromatics mixture charged to the isomerization reactor, usually contains 10 to 40% ethylbenzene. After isomerization, in most cases, the

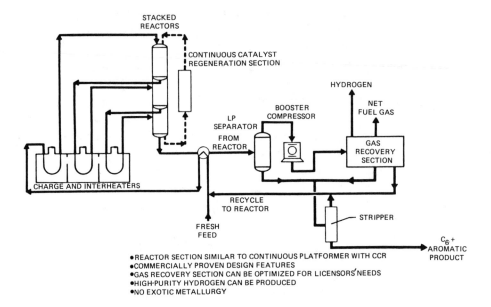

STACKED
REACTORS

CONTINUOUS CATALYST
REGENERATION SECTION

HYDROGEN

NET
FUEL GAS

BOOSTER
COMPRESSOR

LP
SEPARATOR
FROM
REACTOR

GAS
RECOVERY
SECTION

CHARGE AND INTERHEATERS

RECYCLE
TO REACTOR

STRIPPER

FRESH
FEED

C_6+
AROMATIC
PRODUCT

●REACTOR SECTION SIMILAR TO CONTINUOUS PLATFORMER WITH CCR
●COMMERCIALLY PROVEN DESIGN FEATURES
●GAS RECOVERY SECTION CAN BE OPTIMIZED FOR LICENSORS′ NEEDS
●HIGH-PURITY HYDROGEN CAN BE PRODUCED
●NO EXOTIC METALLURGY

Figure 6.1   UOP/BP Cyclar process for LPG aromatization.
(From Anderson et al., 1985.)

p-xylene in the equilibrated mixture is separated by crystalliza-
tion or molecular sieve adsorption, and the remaining liquid is
recycled together with fresh feed to the isomerization reactor.
To avoid the build-up of ethylbenzene concentration in the
recycle loop, the ability of a catalyst to convert ethylbenzene
with a minimum loss of xylenes is of critical importance to a
satisfactory process.

The MVPI process operates by a different reaction pathway
from that of the older processes, such as Octafining. The activity
and stability of the MVPI catalyst and its product selectivity are
the major advantages over the older processes. Octafining uses
a noble metal containing bifunctional catalyst to isomerize
xylenes and convert ethylbenzene to xylenes (Figure 6.3). However, xylene losses are appreciable through other side reactions

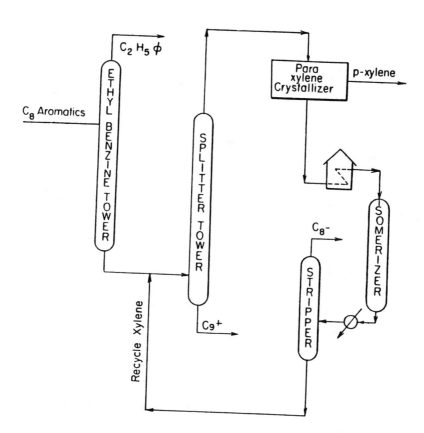

Figure 6.2   Schematic diagram of a xylene isomerization process.   (From Haag and Olson, 1974.)

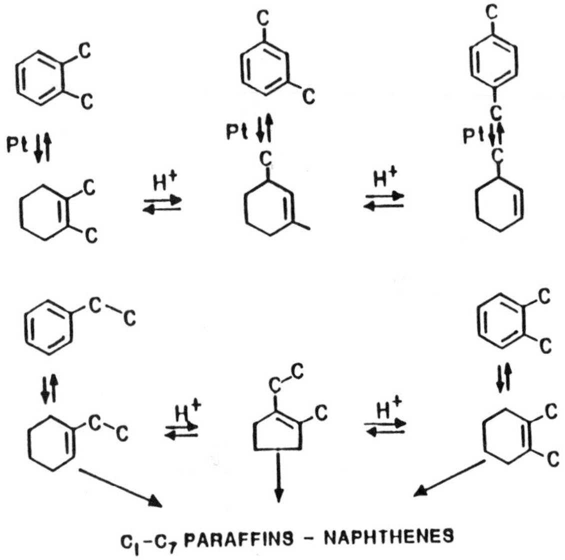

Figure 6.3   Bifunctional isomerization reactions—octafining.
(From Olson and Haag, 1984.)

such as disproportionation and hydrogenolysis. The MVPI
process uses a monofunctional acid catalyst. The process is
carried out at 315–370°C. Ethylbenzene is converted to benzene
and C_{10} aromatics by transalkylation reactions.

Xylene isomerization over acid catalysts is also accompanied
by other side reactions such as disproportionation. Relative to
larger pore zeolites, this side reaction is greatly reduced with the
smaller pore ZSM-5 zeolite. As shown in Figure 6.4, the relative
rate constant of the disproportionation reaction to that of the
isomerization reaction decreases with the decrease in the pore
size of the zeolite, suggesting that the pore size of the zeolite is
the predominant factor influencing selectivity.

With ZSM-5, due to the combined effect of configurational
diffusion and restricted transition state selectivity, the rate of
transalkylation of ethylbenzene is more than 100 times faster

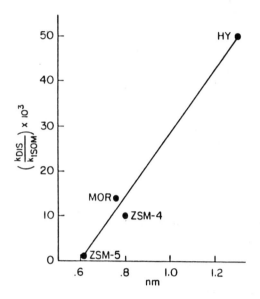

Figure 6.4 Relative rate constants of disproportionation to isomerization of xylenes over various zeolites. (From Olson and Haag, 1974.)

than that of disproportionation of xylenes. Thus, the loss of xylenes is minimized, a major reason for its superb product selectivity in addition to its high activity and stability (Haag and Dwyer, 1979; Olson and Haag, 1984.)

For unextracted feeds which may contain C_8^+ aliphatics, Mobil developed a high temperature xylene isomerization process (MHTI). The catalyst contains a selective hydrogenation function and the process operates at higher temperatures (Olson and Haag, 1984). At these higher temperatures, the aliphatics are cracked to lighter products, and the ethylbenzene is hydrodealkylated to benzene and ethane. Again, xylene losses are exceedingly low. Figure 6.5 shows a schematic diagram of the process.

These new xylene isomerization processes are now in commercial use in more than one-half of the Western World's capacity for p-xylene production.

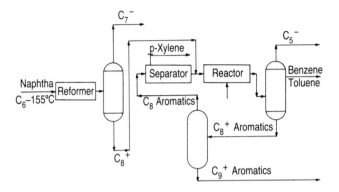

Figure 6.5 High temperature xylene isomerization process. (From Tabak and Morrison, 1980.)

C. Xylene Synthesis (Mobil Toluene Disproportionation Process)

The Mobil toluene disproportionation process (MTDP) is designed to convert toluene to a mixture of xylene isomers and benzene. The rate of toluene disproportionation is about 5000 times slower than that of xylene isomerization (Olson and Haag, 1984). However, at above 450°C, toluene disproportionates readily over ZSM-5 to give a near-equilibrium mixture of benzene and xylenes.

The process employs a fixed bed reactor operating in hydrogen at a total pressure of 20 to 40 atm. Cycle life depends on the ratio of hydrogen to hydrocarbon. At a mole ratio of hydrogen to hydrocarbon of 1.5 to 3, a cycle life of 6 months to 1 year can be expected. Table 6.2 shows a typical operating result.

It was noted in the laboratory and in commercial operation that as the catalyst ages toward the end of cycle, it often undergoes subtle changes to exhibit selective production of *p*-xylene. As shown by the data obtained in a laboratory aging test (Figure 6.6), the *p*-xylene concentration slowly increased with

Table 6.2   Toluene Disproportionation

Operating condition:

Pressure: 33 atm
Temperature: 505°C
H_2/toluene, mol/mol: 2.5
LHSV, vol/vol-hr: 6

Toluene conversion, wt %: 42.6

Product distribution, wt %:

C_5^-	1.5
Benzene	17.9
Toluene	57.4
p-Xylene	4.7
m-Xylene	11.5
o-Xylene	5.2
Ethylbenzene	0.2
C_9^+	1.8

---

on-stream time and eventually exceeded its equilibrium value, reminiscent of the para-selectivity achieved by chemical modification (Chapter 4).

The process has been in commercial use at Mobil's Naples petrochemical complex in Italy since 1975. Three additional units have been licensed worldwide.

Recently Mobil announced a new toluene disproportionation catalyst which operates at 50°C lower temperatures, while giving similar product selectivities (Absil et al., 1988).

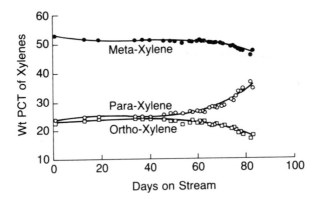

Figure 6.6   Xylene distribution in toluene disproportionation.
(From Nicoletti, 1976.)

II. ETHYLBENZENE AND PARAETHYLTOLUENE

A. Ethylbenzene Synthesis (Mobil/Badger EB Process)

The Mobil/Badger ethylbenzene process produces ethylbenzene,
the feedstock for styrene, by alkylating benzene with ethene
over a ZSM-5 catalyst. This is the first commercial alkylation
process using an acid zeolite catalyst. The process has been in
commercial use since 1976 (Dwyer, 1981) and is gradually
replacing the older processes using the Friedel-Crafts catalysts.
At present, six commercial plants are operating and three more
units have been licensed, bringing the share of the free world
capacity to about 25%.

A simplified process flow diagram is shown in Figure 6.7.
In this process, fresh and recycled benzene is combined with a
diethylbenzene rich stream recovered from the product recovery
section and fed to the alkylation reactor and reacted with fresh
ethene over ZSM-5 to give ethylbenzene. The mole ratio of
benzene to ethene in the feed falls in the range of 5 to 20.
The reaction takes place above 370°C, at 14-27 atm and at a
very-high-weight hourly space velocity of 300-400 (kg total
feed per kg catalyst per hour). As the mole ratio of benzene to

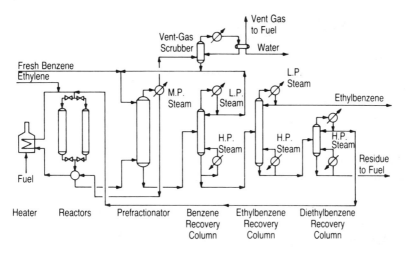

Figure 6.7   New catalyst simplifies Mobil-Badger for making ethylbenzene. (From Dwyer et al., 1976.)

ethene is decreased, single pass conversion of benzene increases at the expense of the production of more diethylbenzene, which must be recovered and recycled. The process has an overall yield of 99.6% ethylbenzene. Heavier residue production has been less than 0.3 wt % relative to ethylbenzene produced (Dwyer et al., 1976; Lewis and Dwyer, 1977).

Because of the low-coke-forming properties of ZSM-5, cycle times of up to 40–60 days between regenerations have been obtained. The high temperature vapor phase alkylation process has major advantages over the low temperature liquid-gas phase process in its energy recovery efficiency and process design simplicity, in addition to the elimination of waste disposal problems associated with Friedel-Crafts catalysts. The process can utilize dilute ethene feed streams.

B. Para-Ethyltoluene Synthesis (PET)

PET has many advantages over styrene (Kaeding et al., 1982); however, *p*-ethyltoluene, the feedstock for *p*-methyl-styrene,

Table 6.3   Alkylation of Toluene with
Ethylene over Modified ZSM-5
Catalysts

Product composition	Wt %
Light gas and benzene	0.9
Toluene	86.2
Ethylbenzene and xylenes	0.5
p-Ethyltoluene	11.9
m-Ethyltoluene	0.4
o-Ethyltoluene	0
C_{10}^+ Aromatics	0.1
Total	100.0
Ethyltoluene isomers (%)	
para	96.7
meta	3.3
ortho	0
Total	100.0

Source: Kaeding et al. (1982).

usually represents about one-third of isomers with the conventional
acidic catalyst, and the problem of separating the para-isomer
from the products of toluene alkylation with ethene has made
it impractical.

Mobil's p-ethyltoluene process is one of the new processes
utilizing the unique configurational diffusion effect of the
channel structure of ZSM-5 to selectively produce para-dialkylbenzenes. Other para-directed processes that are being developed
include the selective toluene disproportionation process (STDP)
(Olson and Haag, 1984), which produces p-xylene in one step
from toluene, and the alkylation of toluene with methanol

process (TAM) (Kaeding et al., 1981), which also produces
p-xylene exclusively.

In the PET process, p-ethyltoluene is produced by the
alkylation of toluene with ethene over a chemically modified
ZSM-5 (Kaeding et al., 1982). The product is 97% p-ethyltoluene (Table 6.3).

REFERENCES

Absil, R. P. L., S. Han, S. M. Leiby, D. O. Marler, J. P.
McWilliams, and D. S. Shihabi, "Toluene Disproportiona-
tion Over ZSM-5 Catalyst," paper presented at the AIChE
Summer Nat. Mtg., Denver, Aug. 21-24, 1988.

Anderson, R. F., J. A. Johnson, and J. R. Mowry, "Cyclar,"
paper presented at the AIChE Spring Natl. Mtg., Houston,
Mar. 24-28, 1985.

Chen, N. Y. and W. E. Garwood, J. Catal. 52, 453 (1978).

Chen, N. Y. and T. Y. Yan, Ind. Eng. Chem., Process Design
Devel. 25, 151 (1986).

Davies, E. E. and A. J. Kolombos, U.S. Pat. 4,180,689, Dec.
25, 1979.

Dwyer, F. G., P. J. Lewis, and F. M. Schneider, Chem. Eng.
83(1), 90 (1976).

Dwyer, F. G., Catalysis of Organic Reactions, W. R. Moser, ed.,
Marcel Dekker, New York, p. 39, 1981.

Johnson, J. A. and G. K. Hilder, "Dehydrocyclodimerization,
Converting LPG to Aromatics," paper presented at the
NPRA Annual Mtg., San Antonio, Tex., Mar. 25-27, 1984.

Garska, D. C. and B. M. Lok, U.S. Pats. 4,499,315-6, Feb. 12,
1985.

Haag, W. O. and D. H. Olson, U.S. Pat. 3,856,871, Dec. 24,
1974.

Haag, W. O. and F. G. Dwyer, "Aromatics Processing with
Intermediate Pore Size Zeolite Catalysts," paper presented
at the AIChE Nat. Mtg., Boston, Mass., Aug. 19, 1979.

Kaeding, W. W., C. Chu, L. B. Young, B. Weinstein, and S. A.
Butter, J. Catal. 67, 159 (1981).

Kaeding, W. W., L. B. Young, and A. G. Prapas, Chemtech 12,
556 (1982).

Kitagawa, H., Y. Sendoda, and Y. Ono, J. Catal. *101*, 12 (1986).

Lewis, P. J. and F. G. Dwyer, Am. Chem. Soc. 173rd Annual
    Mtg., New Orleans, Mar. 1977; Oil Gas J. *75*(40), 55 (1977).

Mole, T. and J. R. Anderson, Appl. Catal. *17*, 141 (1985).

Nicoletti, M. P., laboratory data, 1976.

Oil Gas J. *36*, 31 (1987).

Olson, D. H. and W. O. Haag, Am. Chem. Soc. Symp. Ser. *248*,
    275 (1984).

Tabak, S. A. and R. A. Morrison, U.S. Pat. 4,188,282, Feb. 12,
    1980.

Walsh, D. E. and L. D. Rollmann, J. Catal. *56*, 195 (1979).

7
Applications in Alternate Fuels and Light Olefins

I. METHANOL TO GASOLINE (MTG)

The first MTG process was brought on stream in late 1985 in New Zealand and produces at full capacity 14,500 barrels per day of gasoline. The gasoline produced from methanol contains typically about 32% aromatics, 60% saturates, and 8% olefins and is virtually identical to conventional unleaded gasoline of high research octane number of about 93.

Mobil's methanol-to-gasoline process (MTG) has attracted world-wide attention (Kam et al., 1984). Its discovery has been hailed as the first new route in more than 40 years for the production of gasoline from coal (Meisel, 1981).

The commercial development of this process has been centered on two versions of reactor design, fixed-bed and fluid-bed (Liederman et al., 1980; 1982). The chemistry and the operating characteristics of the conversion process was reviewed by Chang in 1983 (Chang, 1983). In designing these two type of reactors, the major concern is in the control and dissipation of the heat generated by the exothermic conversion reactions.

The first commercial fixed bed MTG unit, the aforementioned 14,500-barrel-per-day gasoline unit, was erected in New Zealand to convert natural gas to gasoline via methanol synthesis. A 100 B/D fluid-bed demonstration plant was built at the Union Rheinishe Braunkohlen Kraftstoff AG (URBK) facility in Wesseling near Bonn, West Germany, in a joint study between Mobil, and URBK and UHDE GmbH, both of Germany, with governmental financial support (Kam et al., 1984). The unit represents a horizontal scale-up of the 4 B/D unit. It has

been in operation since 1982. Testing of the fluid bed design
was completed in 1984.

A. Commercial Fixed-Bed Plant

Shown in Figure 7.1 is the process flow diagram of the New
Zealand plant. The crude methanol from the methanol synthesis
unit is fed at about 21 atm to the MTG unit, which has four
main sections: (1) reactors, (2) product distillation, (3) heavy
gasoline treating to reduce durene concentration, and (4) catalyst
regeneration facilities. The MTG reactors comprise a dehydration reactor following by a parallel train of five conversion
reactors. In the dehydration reactor, methanol is converted over
a dehydration catalyst to an equilibrium mixture of methanol/
dimethyl ether/water, releasing about 20% of the total reaction
heat and raising the temperature of the stream from about
300°C to about 400°C; this mixture is then mixed with 9:1 mol
ratio of recycle gas and converted over ZSM-5 to hydrocarbons.
The dilution by recycle gas controls the reactor exit temperature
to 400°C to 420°C. Heat is recovered from the effluent stream
in a medium pressure steam generator and heat exchanger (Lee
et al., 1980; Penick et al., 1982; Yurchak, 1988).

In actual operation, four of the five conversion reactors
are on stream; the fifth reactor is either in the regeneration
mode or on standby (Fox, 1982). During each operating cycle,
the yield of C_5^+ gasoline slowly increases and its composition
slowly changes from an aromatic-rich gasoline to an olefin-rich
gasoline until methanol breakthrough, at which time the cycle
ends and the catalyst oxidatively regenerates. Typical pilot
plant data on the change in product composition with time on
stream are shown in Table 7.1. Also shown in Table 7.1 is the
yield of 9 RVP (Reid vapor pressure) gasoline, which includes
the alkylate made from the isobutane and light olefinic products.
With four parallel reactors in staggered operation, the composition of the total gasoline product is maintained as a constant
with time.

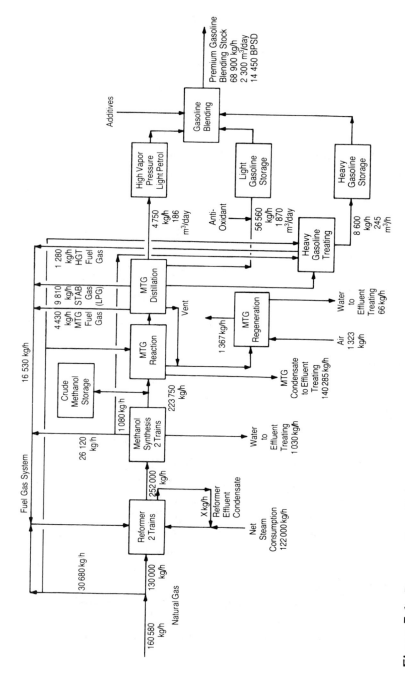

Figure 7.1 Process flow diagram of New Zealand MTG plant. (From Walker, 1985.)

Table 7.1   A Fixed-Bed Aging Study: Hydrocarbon Product
Distribution

	Basis: weight percent				
		Time on stream (h)			
	8	47	95	142	167
Methane, ethane, ethylene	2.4	2.0	2.1	1.5	2.0
Propane	11.0	7.7	7.6	5.6	4.9
n-Butane	5.2	4.0	3.9	2.9	2.6
Isobutane	10.2	9.5	9.5	8.7	8.6
C_3-C_4 Olefins	2.5	2.7	3.4	4.2	3.7
$C_5{}^+$ Nonaromatics	35.0	42.1	42.4	48.3	50.8
Aromatics	33.7	32.0	31.1	28.8	27.4
	100.0	100.0	100.0	100.0	100.0
$C_5{}^+$	68.7	74.1	73.5	77.1	78.2
9 RVP gasoline including alkylate	75.9	81.9	82.8	88.2	88.4
Durene	2.5	4.2	4.3	4.2	4.5

Source: Chang (1983).

It is unique to medium pore zeolites, through which various
tetramethylbenzene isomers diffuse at vastly different rates, that
durene (1,2,4,5-tetramethylbenzene), the smallest among them,
is preferentially formed far above its equilibrium value. Shown
in Table 7.2 is the durene content in the aromatic fraction of a
typical gasoline product from the methanol reactor.

Because of its relatively high melting point (79°C), durene
at sufficiently high concentrations, could crystallize to form a
deposit in the carburetor during cold starts. However, extensive
vehicle performance tests (Fitch and Lee, 1981) showed no
adverse effect due to durene when its concentration is less than

Table 7.2   Durene Content in Aromatics from
Methanol

% A_{10} in aromatics: 8.0	% Durene in A_{10}: 50.1
	Equilibrium at 370°C: 33.4

Source: Chang et al. (1981).

2%. Conventional petroleum derived gasolines in the U.S.
typically contain 0.2 to 0.5% durene.

To reduce the durene concentration in the MTG product,
a heavy gasoline treating process, typical of a mild hydrotreating
process, was developed to process the 177°C+ bottom fraction,
which contains all the durene (Fox, 1982; Penick et al., 1982).
As shown by the data in Table 7.3, the concentration of durene
in the heavy gasoline is reduced by isomerization, dealkylation,
and hydrocracking to less than 15 wt % without suffering any
loss of octane rating. It is interesting to note that among the
isomers only durene has a high melting point (Table 7.4).

The finished gasoline product from a fixed bed MTG process
comprises a mixture of MTG gasoline and varying quantities of
butanes and alkylates depending on volatility requirements and
on the availability of an alkylation unit. The composition and
properties of a typical finished MTG gasoline are shown in
Table 7.5.

B. Fluid-Bed Development

Development of the fluid bed process began with the scale-up of
a bench unit to a 4 B/D pilot plant (Liederman, et al., 1978;
Kam and Lee, 1978). Fluidized beds offer the advantage over
fixed beds in the easiness of temperature control and heat
removal, but are complicated by the necessity that the internals
of the reactor must be designed to achieve complete conversion
of methanol in order to avoid the costly step of recovering

Table 7.3   Heavy Gasoline Treating Yields (Wt %)

	Feed	Product
C_2^-	—	0.3
$C_3 + C_4$	—	1.96
C_5^+ P + N	0.65	4.52
C_6 Aromatics	0.00	0.03
C_7 Aromatics	0.00	0.52
C_8 Aromatics	0.74	5.74
C_9 Aromatics	23.04	29.41
Durene, 1, 2, 4, 5-tetramethylbenzene	43.69	13.62
Isodurene, 1, 2, 3, 5-tetramethylbenzene	8.13	16.06
Prehnitene, 1, 2, 3, 4-tetramethylbenzene	2.80	3.60
Other C_{10}^+ aromatics	20.95	24.24

Source: Yurchak (1988).

Table 7.4   Melting Point of Tetramethyl Benzenes

	Melting point, °C
Prehnitene, 1, 2, 3, 4-tetramethylbenzene	−23.7
Isodurene, 1, 2, 3, 5-tetramethylbenzene	−6.3
Durene, 1, 2, 4, 5-tetramethylbenzene	+79.2

Source: Rossini et al. (1953).

Table 7.5   Typical Finished Gasoline
Properties

Components, vol %	
C_5^+ Gasoline	95
C_4's	5
Octane number	
Clear research	93
Clear motor	83
Distillation (D-86), RC (RF)	
10%	46 (115)
50%	99 (210)
90%	166 (330)
EP	204 (400)
Existent gum, mg/100 ml	1
Sulfur, nitrogen, ppm	<10
Composition, wt %	
n-Paraffins	4.6
i-Paraffins	41.6
Olefins	9.5
Naphthenes	9.2
Aromatics	35.1
Durene	1.84

Source: Yurchak (1988).

unconverted methanol from the reactor effluent. The fluid-bed unit operates at much lower pressures than the fixed-bed unit and its yield pattern is different from that of the fixed bed: lower C_5^+ liquid yield and higher-light-olefins yield. Table 7.6 compares the process conditions and product yields from these two systems. The composition and properties of a typical finished gasoline from a fluid-bed pilot plant are shown in Table 7.7.

Performance of the bench unit and the 4 B/D pilot plant was recently reviewed by Chang (1983).

A schematic diagram of the 100 B/D demonstration plant is shown in Figure 7.2. The reactor consists of a dense fluid-bed section located above a dilute phase riser. Reaction heat is removed either by circulating the catalyst through an external cooler or through heat exchangers placed within the catalyst bed to produce steam at pressures up to 100 atm (Penick et al., 1982; Gierlich et al., 1985).

Table 7.8 lists the operating variables of the fluid-bed demonstration plant.

One distinct advantage of the fluid-bed reactor over the fixed-bed reactor is its ability to maintain constant catalyst activity by continuous regeneration and thereby produce a constant quality product. By adjusting the operating variables, i.e., temperature and pressure, it is possible to maximize either the gasoline yield or its octane rating as shown by the data in Figures 7.3 and 7.4.

Table 7.6 Process Conditions and Product Yields from MTG Processes

	Fixed bed	Fluid bed
Conditions		
Methanol/water chg. (W/W)	83/17	82/17
Dehydration reactor inlet temperature (°C)	316	

Table 7.6　(Continued)

	Fixed bed	Fluid bed
Conditions (continued)		
Dehydration reactor outlet temperature ($^\circ$C)	304	—
Conversion reactor inlet temperature ($^\circ$C)	360	413
Conversion reactor outlet temperature ($^\circ$C)	415	413
Pressure (kPa)	2,170	275
Recycle ratio (mol/mol chg.)	9.0	—
Space velocity (WHSV)	2.0	1.0
Yields (Wt % of methanol charged)		
Methanol + ether	0.0	0.2
Hydrocarbons	43.4	43.5
Water	56.0	56.0
CO, CO_2	0.2	0.2
	100.0	100.0
Hydrocarbon product (wt %)		
Light gas	1.4	5.6
Propane	5.5	5.9
Propylene	0.2	5.0
Isobutane	8.6	14.5
n-Butane	3.3	1.7
Butenes	1.1	7.3
C_5^+ Gasoline	79.9	60.0
	100.0	100.0
Gasoline (including alkylate) [RVP-62 kPa (9 psi)]	85.0	88.0
LPG	13.6	6.4
Fuel gas	1.4	5.6
	100.0	100.0

Source: Penick et al. (1982).

Table 7.7   Typical Properties of Finished
Gasoline from Fluid Bed MTG Unit

Components, wt %		
Butanes	3.2	
Alkylates	28.6	
C_5^+ Gasoline	68.2	
	100.0	
Composition, wt %		
Paraffins	56	
Olefins	7	
Naphthenes	4	
Aromatics	33	
	100	
Octane		
	Research	Motor
Clear	96.8	87.4
Reid vapor pressure	9	
Specific gravity	0.730	
Sulfur, wt %	Nil	
Nitrogen, wt %	Nil	
Durene, wt %	3.8	
Corrosion, copper strip	1A	
ASTM Distillation, RC		
10%	47	
30%	70	
50%	103	
90%	169	

Source: Lee et al., 1979.

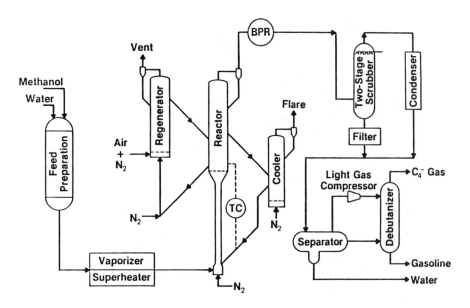

Figure 7.2   100 B/D fluid MTG plant.   (From Penick et al., 1982.)

Table 7.8   Operating Variables of Fluid Bed Demonstration Plant

Reactor pressure	2.7–4.5 bar abs.
Reactor temperature	380–430°C
WHSV	0.5–1.75 H^{-1}
Methanol feed rate	500–1050 kg/h
Methanol conversion	>99.9%
Time on stream	8600 hrs
Methanol processed	6800 metric tons
Max. gasoline yield (incl. alkylate)	92%
Service factor total	65%
Service factor during scheduled test runs	99%

Source: Gierlich et al. (1985).

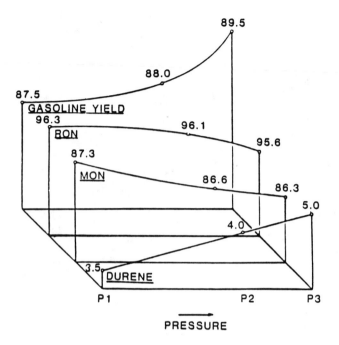

Figure 7.3   Effect of pressure on gasoline quality. (From Gierlich et al., 1985.)

Data obtained in the demonstration plant will serve the basis for a conceptual design of a commercial size fluid-bed reactor, processing 2,500 tonnes/day (18,000 B/D) of methanol (Gierlich et al., 1985). The design was completed in 1985 and is available for commercial license.

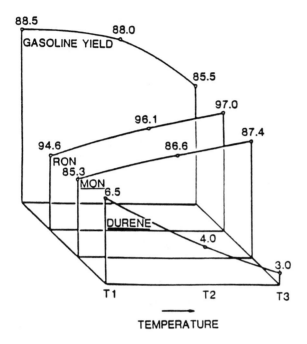

Figure 7.4   Effect of temperature on gasoline quality. (From Gierlich et al., 1985.)

II. METHANOL TO LIGHT OLEFINS (MTO)

Instead of producing an aromatic gasoline, methanol can also be converted to yield mostly light olefins.

Mobil's methanol-to-olefins process (MTO) is based on a discovery that by reducing the acidity of the zeolite and raising the operating temperature to above 500°C, the relative rate of olefin formation reactions and aromatization reactions are different enough so that C_2 to C_5 olefin yield as high as 80% has been obtained (Chang et al., 1984).

These laboratory results have been confirmed in a 4 B/D fluid-bed pilot plant (Gould et al., 1986). Shown in Table 7.9 is a comparison of the results obtained in these two units.

Table 7.9   Comparison of Bench Scale and Pilot Plant Data on Methanol-to-Olefin Process

Reaction conditions: Pressure: 1 atm, temperature: 482°C		
	Wt %	
	Bench unit	Pilot plant
Methanol conversion	100	100
Product yields		
Ethene	5.8	5.2
Propene	33.7	32.9
Butenes	18.4	19.1
Pentenes	14.8	12.0
C_6^+ Olefins	5.5	7.7
C_3^- Paraffins	5.0	5.0
C_4^+ Paraffins/naphthenes	10.3	11.6
Aromatics	6.5	6.5
	100.0	100.0
Total olefins	78.2	76.9
Total C_4^+ gasoline	16.8	18.1
Light paraffins	5.0	5.0

Source: Gould et al. (1986).

The process has also been successfully demonstrated in the 100-barrel-per-day fluid-bed MTG unit in Germany (Gould et al., 1986). The 100 B/D reactor is 60 cm in diameter and 12 m high, a modified version of the MTG reactor. The reactor is equipped with heat exchanger coils immersed in the dense fluid bed. Because the unit was originally designed for the MTG process, the reactor could not be operated at near-atmospheric pressure, hence the demonstration runs were carried out at 2.5 atm and 500°C. As expected, higher operating pressure reduces the olefin yields to 61% and increases the gasoline yield to 32% (Table 7.10). Considering the excellent reproducibility between the pilot plant and the demonstration unit, the scale-up study was considered a success.

The MTO process may be coupled with Mobil's Olefins to Gasoline and Distillate Process (MOGD) described earlier, to provide another synthetic fuel process for producing gasoline and distillates from methanol and/or coal. A schematic diagram of the combined process is shown in Figure 7.5. Methanol is fed to the MTO reactor where it is converted to hydrocarbons and water. The hydrocarbons are separated into light olefins, an ethene-rich stream and an aromatic gasoline stream. The light olefins are fed to the MOGD unit along with an MOGD recycle stream. The MOGD product is then separated into raw distillate, gasoline, and a small fuel gas and LPG streams. Hydrotreating the raw distillate gives the final distillate product.

When the products of the MTO/MOGD process are compared with that of Fischer-Tropsch synthesis, it is noted that in the MTO process the yield of C_3^- paraffins is typically less than 5 wt % of total hydrocarbons. This is significantly less than that produced by the Fischer-Tropsch synthesis (typically more than 15%). Thus it is possible to produce liquid yields in excess of 90% of the total hydrocarbon products by this new route.

Since both MTO and MOGD have flexible product slates, this results in a combined process that should be able to meet the variable seasonal market demand of gasoline or distillate. Table 7.11 presents an estimate of product yields from this combined process.

Table 7.10   Comparison of Pilot Plant and
Demonstration Unit Data on Methanol-to-Olefin
Process

Reaction conditions: Pressure: 2.5 atm, temperature: 500°C		
	Wt %	
	Pilot plant	Demonstration unit
Methanol conversion	100	100
Product yields		
C_3^- Paraffins	7	7
Olefins	63	61
Total C_4^+ gasoline	30	32
	100	100

Source: Gould et al. (1986).

Table 7.11   MTO/MOGD Product Yields

Distillate/gasoline ratio	0.65	0.95
Product yield, wt %		
LPG	5.1	5.1
Gasoline	57.0	48.0
Diesel	37.9	46.9
	100	100

Source: Tabak et al. (1986).

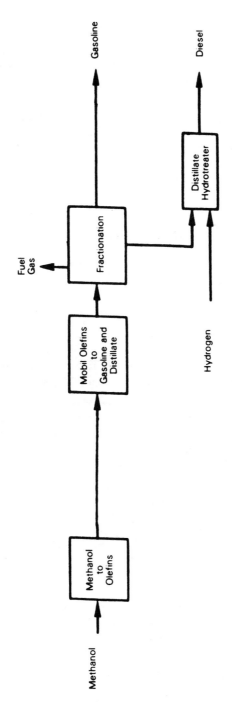

Figure 7.5 An example of an integrated MTO/MOGD process (schematic). (From Tabak et al., 1986.)

It also provides a means to convert methanol to high quality jet and diesel fuels.

More recently, Union Carbide (1986) announced the development of a fluid-bed methanol-to-olefin process using a silico-aluminophosphate (SAPO) catalyst. The SAPO catalyst reportedly produces olefins in very high yields (>90%) and the process can be modified to produce 60% ethene or propene (Kaiser, 1985; Pellet et al., 1987).

III. SYNTHESIS GAS CONVERSION

A. Composite Catalysts for Direct Synthesis Gas Conversion

While zeolites do not have sufficient intrinsic catalytic activity for CO reduction, they may be combined with carbon monoxide reduction catalysts such as methanol synthesis catalysts (Chang et al., 1984; Fujimoto et al., 1984) or iron Fischer-Tropsch catalyst (Chang et al., 1979; Caesar et al., 1979; Dwyer and Garwood, 1984) and a ruthenium Fischer-Tropsch catalyst (Huang and Haag, 1981a; Chen et al., 1984) to alter the distribution of the final product. Fischer-Tropsch synthesis typically produces a mixture of paraffins and olefins ranging from methane to high-molecular-weight waxes. The formation of C_5^+ hydrocarbons is usually favored by low reaction temperatures. By mixing the conventional Fischer-Tropsch catalyst with medium pore molecular sieves, such as HZSM-5, (Caesar et al., 1979) or SAPO-11 (Coughlin and Rabo, 1985; Coughlin, 1986), it has been demonstrated in the laboratory that the boiling range of the hydrocarbon product can be narrowed and aromatic liquid products boiling in the gasoline range have been made in a single step directly from the synthesis gas. The olefin isomerization activity is also enhanced by the presence of the molecular sieve component to reduce the pour point of the liquid product. In the absence of the molecular sieve component, the liquid product comprises essentially waxy normal olefins and paraffins. The

acidity of the SiO_2/Al_2O_3 ratio of the ZSM-5 in an intimate
mixture of ZSM-5 and Fe(K) catalysts was found to have a
significant effect on the olefinicity of the hydrocarbon products.
Olefin contents increased with increasing SiO_2/Al_2O_3 ratios
(Dwyer and Garwood, 1984) as shown by the data in Figure
7.6. The degree of intimacy of mixing of these two components
in the composite catalyst was also found to affect the composition of the liquid product (Huang and Haag, 1981a). For
example, Table 7.12 shows that without ZSM-5 the ruthenium
Fischer-Tropsch catalyst makes only nonaromatic product.
Adding ZSM-5 to the composite catalyst as a physical mixture
changes the composition of the liquid product to that of an
aromatic gasoline. Using a more intimate mixture, prepared by
impregnating ZSM-5 with $RuCl_3 \cdot 3H_2O$, promotes the aromatics
alkylation reaction and shifts the products to $C_{11}{}^+$ heavy
aromatics.

One of the disadvantages in using these composite catalysts
for the direct conversion of synthesis gas is in situations where
maximizing $C_5{}^+$ product is desired. The problem lies in a mis-
match of the reaction temperature required for optimal performance and is illustrated by the data shown in Figure 7.7,
obtained with a physical mixture of RuO_2 and HZSM-5 over
a temperature range of 260°C to 320°C. Because the zeolitic
component shows little catalytic activity below 260°C, while
$C_4{}^-$ gas increases rapidly with temperatures beyond 260°C with
known Fischer-Tropsch catalysts, it is difficult to select a
temperature at which the zeolitic component is effective without
excess gas production.

B. Mobil Two-Stage Slurry Fischer-Tropsch/ ZSM-5 Process

Low methane formation and high aromatic gasoline production
can be achieved when the Fischer-Tropsch catalyst and the
zeolite catalyst are placed in separate reactors and operated in a
cascade mode under different reaction conditions (Huang and
Haag, 1981b).

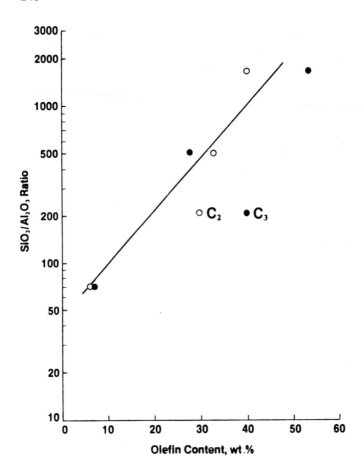

Figure 7.6   Effect of SiO_2/Al_2O_3 ratio on C_2 and C_3 olefin content. (From Dwyer and Garwood, 1984.)

Table 7.12   Effect of Intimacy of Mixing on Product
Distribution

Reaction conditions:
Pressure: 51 atm, temperature: 294–304°C, H_2/CO: 2 mol/mol

		Catalyst	
	Ru/Al_2O_3	Ru/ZSM-5	Ru/Al_2O_3 + ZSM-5
Ru loading, wt %	0.5	5	5
Conversion, mol %	94	86	99
Reactor effluent, wt %			
Hydrocarbons	37	38	40
H_2	0	1	0
CO	11	15	0
CO_2	6	2	20
H_2O	46	44	40
Hydrocarbon composition, wt %			
$C_1 + C_2$	33	38	43
$C_3 + C_4$	8	16	14
C_5^+	59	46	43
Aromatics in C_5^+, wt %	0	25	24
Aromatics distribution, wt %			
Benzene-A_{10}	0	79	97
A_{11}^+	0	21	3

Source: Huang and Haag (1981a).

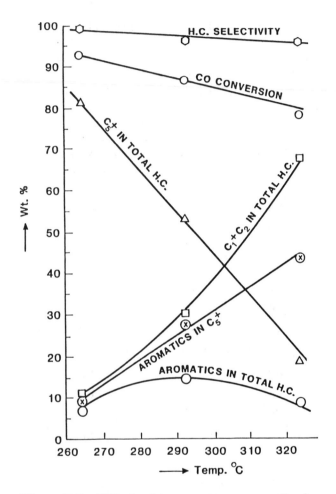

Figure 7.7   Effect of temperature on synthesis gas conversion over 5% Ru(as RuO_2)/ZSM-5 (294°C, GHSV = 480 and H_2/CO = 2/1).   (From Huang and Haag, 1981b.)

This is the basis of Mobil's two-stage slurry Fischer-Tropsch/ZSM-5 process currently under development. The process combines the slurry-phase Fischer-Tropsch synthesis technology with a fixed-bed ZSM-5 reactor, which converts the vapor phase product from the first stage reactor to high quality gasoline (Kuo, 1983). To minimize the yield of methane and ethane, the operation mode of the slurry reactor leads to the production of C_{20}^+ waxy products that accumulate in the reactor and is removed for further upgrading (Kuo, 1985).

Figure 7.8 shows a simplified flow diagram of the two-stage pilot plant. The slurry Fischer-Tropsch reactor, shown in Figure 7.9, measures 5 cm I.D., and the liquid depth may be varied between 305 cm and 760 cm. It was designed to hold about 1600 g of catalyst and process up to 3 m^3(STP)/hr of synthesis gas. Reaction heat is removed from the reactor by circulating a hydrocarbon based coolant through an outer jacket. The second stage comprises two ZSM-5 reactors. Each consists of a fixed catalyst bed, 5 cm in diameter and 20 cm long. One of them is in operation while the second reactor is either in the regeneration mode or on standby.

Typical pilot plant data on the first-stage reactor feeding a syngas containing 0.68 mole ratio of hydrogen to CO are presented in Table 7.13. The overhead effluent from the reactor was processed in the second-stage reactor to the final product (Table 7.14). Similar to the fixed-bed MTG reactor, during each operating cycle, the catalyst slowly deactivated and the composition of C_5^+ gasoline slowly changed from an aromatic-rich gasoline to an olefin-rich gasoline. To maintain constant quality product, the reactor temperature was gradually raised from the start-of-cycle temperature of 300°C to the end-of-cycle temperature of about 410°C over a period of 30 days, at which time the catalyst was oxidatively regenerated. The average gasoline contains about 45 wt % of aromatics, 15 wt % olefins and 40 wt % paraffins and has a research clear octane (R+O) of about 90.

The slurry wax removed from the first-stage reactor contains catalyst fines. After settling to remove most of the

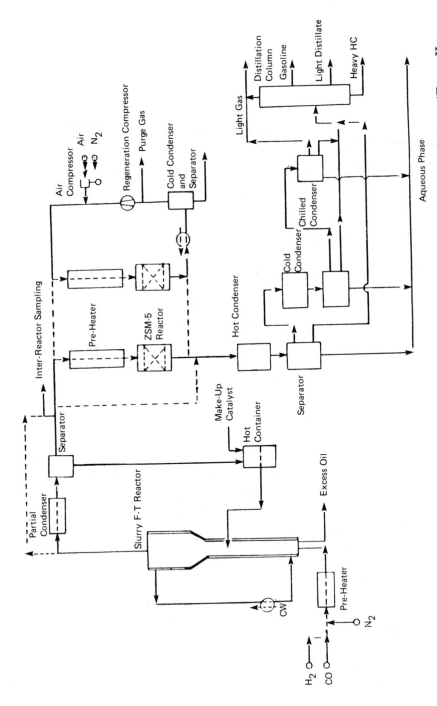

Figure 7.8 Simplified flow diagram of two-stage pilot plant for synthesis gas conversion. (From Kuo, 1983.)

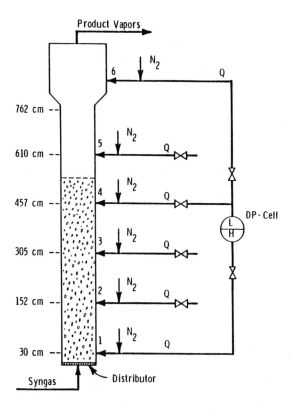

Q: DP-Cell Lines

Figure 7.9 Schematic arrangement of DP-cell for liquid level measurement. (From Kuo, 1983.)

Table 7.13   Typical Pilot Plant Data on First-Stage Slurry Fischer-Tropsch Reactor

---

Operation conditions:

Temperature:	$260°C$
Pressure:	15 atm
Space velocity:	2 1/g Fe-hr
Catalyst:	$Fe/Cu/K_2CO_3$

Catalyst loading: 19.5 wt %

Syngas conversion, mol %

Hydrogen	78.8
CO	88.7
H_2 + CO	84.7

Yields, wt % of products

Hydrocarbons (including oxygenates)	22.0
CO_2	66.4
H_2O	0.8
Hydrogen	0.9
CO	9.9

Composition of hydrocarbons, wt %

Methane	7.9
Ethene	1.5
Ethane	3.0
Propene	8.2
Propane	1.9
Butenes	6.0
Butanes	2.0
C_5^+ overhead	54.3
Slurry wax	6.2

Source: Kuo (1983).

Table 7.14   Typical Pilot Plant
Data on Second-Stage ZSM-5
Reactor

Operation conditions:	
Temperature:	320°C
Pressure:	13 atm

Overall yields, wt %	
Methane	7.3
Ethene	1.0
Ethane	3.3
Propene	0.9
Propane	8.5
Butenes	1.4
Butanes	19.0
C_5–C_{11} Gasoline	49.3
C_{12}^+ and slurry wax	9.3
	100.0

Source: Kuo (1985).

catalyst, it is vacuum-distilled to recover a solids-free wax
stream and the still bottom is thermally cracked and recycled
to the vacuum tower. The wax stream can be upgraded to
gasoline and distillates by conventional hydrocracking/cracking
processes.

Based on the pilot plant study, a conceptual design for a
27,000 B/D gasoline plant has been developed (Kuo, 1985).

REFERENCES

Caesar, P. D., J. B. Brennan, W. E. Garwood, and J. Ciric, J.
Catal. 56, 274 (1979).

Chang, C. D., W. H. Lang, and A. J. Silvestri, J. Catal. *56*, 268 (1979).

Chang, C. D., W. H. Lang, and W. K. Bell, *Catalysis of Organic Reactions*, W. R. Mosher, ed., Marcel Dekker, New York, p. 73, 1981.

Chang, C. D., Catal. Rev.-Sci. Eng. *25*, 1 (1983); *Hydrocarbons from Methanol*, Marcel Dekker, New York, 1983.

Chang, C. D., C. T.-W. Chu, and R. F. Socha, J. Catal. *86*, 289 (1984).

Chang, C. D., J. N. Miale, and R. F. Socha, J. Catal. *90*, 84 (1984).

Chen, Y. W., H. T. Wang, and J. G. Goodwin, J. Catal. *85*, 499 (1984).

Coughlin, P. K., and J. A. Rabo, U.S. Pat. 4,556,645, Dec. 3, 1985.

Coughlin, P. K., U.S. Pat. 4,579,830, Apr. 1, 1986; U.S. Pat. 4,632,941, Dec. 12, 1986.

Dwyer, F. G. and W. E. Garwood, "Catalytic Conversions of Synthesis Gas and Alcohols to Chemicals," R. G. Herman, ed., Plenum, New York, p. 167, (1984).

Fitch, F. B. and W. Lee, "Methanol-to-Gasoline, An Alternative Route to High-Quality Gasoline," paper presented at the Int. Pacific Conf. on Automotive Eng., SAE Tech. Paper 811403, Honolulu, Nov. 16–19, 1981.

Fox, J. M., "The Fixed-Bed Methanol-to-Gasoline Process Proposed for New Zealand," paper presented at the Coal Gasification Conf., Australian Inst. Petro., Adelaide, Australia, Mar. 2, 1982.

Fujimoto, K., K. Yoshihiro, and H. Tominaga, J. Catal. *87*, 101 (1984).

Gierlich, H. H., K. H. Keim, N. Thiagarajan, E. Nitschke, A. Y. Kam, and N. Daviduk, "Successful Scale-Up of the Third Bed Methanol to Gasoline (MTG) Process to 100 BPD Demonstration Plant," paper presented at the 2nd EPRI Conf. "Synthetic Fuels—Status and Directions," San Francisco, Apr. 15–19, 1985.

Gould, R. M., A. A. Avidan, J. L. Soto, C. D. Chang, and R. F. Socha, "Scale-Up of a Fluid-Bed Process for Production of Light Olefins from Methanol," paper presented at the AIChE Nat. Mtg., New Orleans, Apr. 6–10, 1986.

Huang, T. J. and W. O. Haag, ACS Symp. Ser. *152*, 307 (1981a).

Huang, T. J. and W. O. Haag, "Aromatic Gasoline Production from Synthesis Gas via a Two-Stage Process," paper presented at the 91st AIChE Nat. Mtg., Detroit, Aug. 16–19 (1981b).

Kaiser, S. W., U.S. Pat. 4,499,327, Feb. 12, 1985; U.S. Pat. 4,524,234, June 18, 1985.

Kam, A. Y. and W. Lee, "Fluid-Bed Process Studies on Selective Conversion of Methanol to High Octane Gasoline," Final Report, DOE Contract EX-76-C-01-2490, U.S. Dept. of Energy, Apr. 1978.

Kam, A. Y., M. Schreiner, and S. Yurchak, "Handbook of Synfuels Technology," R. A. Meyers, ed., McGraw-Hill, New York, pp. 2–75, 1984.

Kuo, J. C. W., "Slurry Fischer-Tropsch/Mobil Two Stage Process of Converting Syngas to High Octane Gasoline," Final Report, DOE Contract DE-AC22-80PC30022, June 1983.

Kuo, J. C. W., "Two Stage Process for Conversion of Synthesis Gas to High Quality Transportation Fuels," Final Report, Dept. Energy Contract DE-AC22-83PC60019, Oct. 1985.

Lee, W., J. Maziuk, V. W. Weekman, and S. Yurchak, *Large Chemical Plants*, G. F. Froment, ed., Elsevier, Amsterdam, p. 171, 1979.

Lee, W., S. Yurchak, D. Daviduk, and J. Maziuk, "A Fixed-Bed Process for Methanol-to-Gasoline Conversion," Paper AM-80-41, NPRA Mtg., New Orleans, Mar. 23–25, 1980.

Liederman, D., S. M. Jacob, S. E. Voltz, and J. J. Wise, Ind. Eng. Chem., Process Des. Dev. *17*, 340 (1978).

Liederman, D., S. Yurchak, J. C. W. Kuo, and W. Lee, "Mobil Methanol-to-Gasoline Process," paper presented at the 15th Intersoc. Energy Conv. Eng. Conf., Seattle, Aug. 1980.

Liederman, D., S. Yurchak, J. C. W. Kuo, and W. Lee, J. Energy *6*, 340 (1982).

Meisel, S. L., Philos. Trans. Roy. Soc. London *A300*, 157 (1981).

Pellet, R. J., J. A. Rabo, G. N. Long, and P. K. Coughlin, "Reactions of C_8 Aromatics Catalyzed by Aluminophosphate Based Molecular Sieves," Paper B-2, 10th North Am. Mtg. Catal. Soc., San Diego, May 17–22, 1987; R. J. Pellet,

J. A. Rabo, G. N. Long, "Olefin Oligomerization Catalyzed by Aluminophosphate Based Molecular Sieves," Poster Paper 75, 10th North Am. Mtg. Catal. Soc., San Diego, May 17-22, 1987.

Penick, J. E., W. Lee, and J. Maziuk, "Development of the Methanol-to-Gasoline (MTG) Process," paper presented at the Intern. Symp. Chem. Reactor Eng. (ISCRE-7) Boston, Oct. 4, 1982.

Rossini, F. D., K. S. Pitzer, R. L. Arnett, R. M. Braun, and G. C. Pimental, "Selected Values of Physical and Thermodynamics Properties of Hydrocarbons and Related Compounds," API Research Project 44, Carnegie Press, Pittsburgh, 1953.

Tabak, S. A., A. A. Avidan, and F. J. Krambeck, "Production of Synthetic Gasoline and Diesel Fuels from Non-Petroleum Sources," paper presented at the ACS Nat. Mtg., New York, Apr. 13-15, 1986.

Union Carbide Corp., Oil Gas J., 31, Dec. 22, 1986.

Walker, B. V., "Synthetic and Alternative Fuels Production in New Zealand," paper presented at EPRI Conf. on Coal Gasification and Synthetic Fuels for Power Generation, San Francisco, Apr. 14-19, 1985.

Yurchak, S., Stud. Surf. Sci. Catal. 36, 251 (1988).

8
New Opportunities in Shape Selective Catalysis

I. UPGRADING WAXY CRUDES

In principle, shape selective cracking is applicable to the processing of waxy crudes to improve their fluidity and eliminate the need to install heated pipelines to transport these crudes. Laboratory experiments have demonstrated that paraffinic waxy crude oils from various parts of the world, including the U.S. (Utah), Indonesia, Australia, North Africa, China and Newfoundland (Chen and Shihabi, 1981; Shihabi, 1981), can be processed in a simple fixed-bed reactor with or without the presence of hydrogen to improve their low temperature fluidity and at the same time increase the quantity and quality of their distillate fraction.

Figure 8.1 illustrates the upgrading of a barrel of Daqing (Taching) whole crude into low-pour products. The con-

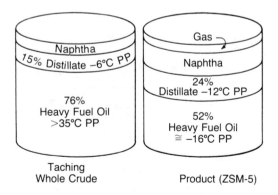

Figure 8.1 Opportunities for direct catalytic upgrading of waxy taching whole crude. (From Wise et al., 1986.)

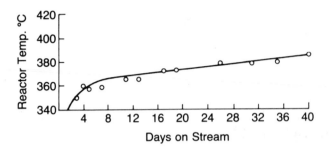

Figure 8.2   Daqing (taching) crude temperature-adjusted to
−12°C pour point—LHSV = 1, pressure = 22 atm; no gas circulation. (From Shihabi, 1981.)

version reduces the heavy oil fraction from 76% to 52% and
decreases the pour point of that fraction to −16°C (Wise et al.,
1986).

Limited long term aging studies indicate that under a
moderate total pressure of 15 to 30 atm, catalyst life exceeds
one month in the absence of hydrogen (Figure 8.2). Catalyst
activity can be restored by water (Shihabi, 1983). However, the
concept requires extensive development study because there is
scarecely any practical experience in on-site catalytic processing.

II. UPGRADING SHALE OILS

Synthetic crudes derived from shale oils typically possess pour
points on the order of 25 to 35°C, too high to be marketed as
distillate fuels because of their poor low temperature fluidity
characteristics (Walsh, 1984; Angevine et al., 1984; LaPierre et
al., 1986).

Table 8.1 shows some typical properties of shale oils. It
is noted that in addition to their high pour points, shale oils are
also high in nitrogen and trace metals and generally require a
hydrogenative pretreatment step before catalytic upgrading
(Miller et al., 1982).

Table 8.1   Physical Properties of Shale Oils

	Analyses, wt %	
	In situ (Occidental)	Retort (Paraho)
Hydrogen	11.92	11.24
Nitrogen	1.45	1.86
Sulfur	0.60	0.71
Oxygen	0.90	1.30
Nickel	0.0012	0.00016
Iron	0.0065	0.0095
Arsenic	0.0028	0.0033
Bromine no.	26.1	43.9
API gravity	23.4	20.5
Pour point °C	16	27
Distillation (D1160) °C		
IBP	189	208
10%	253	245
50%	370	378
90%	512	474

Source: LaPierre et al. (1986).

Catalysts prepared from an admixture of a medium pore zeolite and a porous refractory oxide supported with VIB and Group VIII metals have been designed to reduce pour point, nitrogen, and sulfur with minimum consumption of hydrogen (Ward and Carlson, 1986). Alternatively, the shale oil may be processed in a cascaded two-stage process (Gorring and Smith, 1979; LaPierre et al., 1986) that combines a first-stage hydrotreating reactor with a second-stage shape selective dewaxing reactor to yield a low-pour-point synthetic crude with minimum

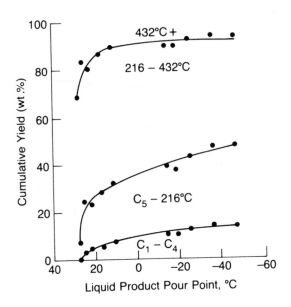

Figure 8.3   Hydrotreating/dewaxing yields—Paraho shale oil.
(From LaPierre et al., 1986.)

additional hydrogen consumption over the hydrotreating require-
ment. Results obtained with Paraho shale oil is shown in
Figure 8.3. The product distribution was found to be dependent
on the pour point of the liquid product. The yield of naphtha
and LPG increased as the pour point of the liquid product
decreased, the result of selective cracking of the paraffinic fraction of the feed.
     The activity of the dewaxing catalyst at 425°C and above
was found to be strongly dependent on the basic nitrogen content of the first-stage product and not affected by the high concentration of gaseous ammonia cascaded from the first-stage
reactor to the dewaxing reactor, suggesting that the catalyst
may function in this case as a nonacidic or low acidity catalyst.

III. DEWAXING HYDROGEN DONOR SOLVENT
FOR COAL LIQUEFACTION

Recent progress made by the use of hydrogen donor solvent for
coal liquefaction through short-contact-time thermal reactions
improves selectivity in the conversion of coal to higher quality
liquids, and the conservation of hydrogen, which is costly to
produce (Whitehurst, 1980; Whitehurst et al., 1984). The effectiveness of a molecule as a hydrogen transfer agent depends
strongly on the chemical structure of the molecule. Labile
hydrogen of the hydroaromatics in the solvent is particularly
effective in transferring hydrogen to stabilize the thermally
cracked radical fragments. Thus, 8 hydrogen atoms per molecule
can be transferred from octahydrophenanthrene, while octa-
decane, a saturated hydrocarbon, has essentially no transferrable
hydrogen (Aiura et al., 1984).

In order to maintain the effectiveness of a hydrogen donor
solvent, the build-up of n-paraffins in the recycle solvent in some
of the coal liquefaction processes must be selectively removed
(Yao et al., 1986a; 1986b). A process known as the Nedol
process, incorporating such concepts, is being developed in
Australia and Japan under the Japanese New Energy Development Organization (NEDO) (Pace Synthetic Fuels Report, 1987)
for the liquefaction of Victoria Brown coal and bituminous coal.
Among the various processing schemes is the use of ZSM-5 for
the dewaxing of the recycle solvent (Mitarai, 1987).

IV. CONVERTING NATURAL GAS TO
LIQUID HYDROCARBONS

Natural gas, over 3 quadrillion (10^{15}) cubic feet in estimated
proven reserves worldwide (McCaslin, 1986) is another source

of energy and chemical feedstock. At present, only a tiny fraction of the natural gas consumed is upgraded to more valuable products such as methanol and ammonia. Converting natural gas to liquid transportation fuels could certainly become an important industry in the future as the supply of crude oils dwindles and the price differential between natural gas and the liquid fuels justifies it.

The synthesis of methanol from natural gas coupled with Mobil's methanol-to-gasoline process (MTG) is one of the newest technologies for the conversion of methane to liquid hydrocarbons using shape selective zeolites as the catalyst.

Another approach involves the direct conversion of natural gas to hydrocarbons by an oxidative coupling reaction (Sofranko, et al., 1987). In this approach, methane is converted to higher-molecular-weight hydrocarbons by contacting it with a metal oxide in a cyclic process. In the methane reactor, the metal oxide is reduced while methane is converted to C_2^+ hydrocarbons and such by-products as water, carbon oxides, and coke. The reduced oxide is then reoxidized in a separate reactor and circulated back to the methane reactor. The C_2^+ product can be upgraded to gasoline and distillates by coupling the methane reactor with Mobil's MOGD process (Haxbun, 1985; Jones et al., 1986), again using shape selective zeolites as the catalyst.

The critical step in these approaches is the activation of methane, whether by high temperature steam or by oxygen. Other approaches to the activation process have also been studied. For example, Anderson and Tsai (1985) studied the activation of methane over HZSM-5 using nitrous oxide as the oxidant. As discussed in Chapter 4, methane can also be activated by reacting with halogen and sulfur. The resulting alkyl halides and mercaptans can be converted to higher-molecular-weight hydrocarbons over the medium pore molecular sieve catalysts (Chang et al., 1975a; Butter et al., 1975; Kaiser, 1987; Brophy et al., 1987). The difficulty with these latter approaches lies in the costly step of recovering the oxidants.

V. PRE-ENGINE CONVERTER

From the point of view of conserving global energy consumption, efforts to improve engine performance by using higher octane fuels could be counterbalanced by the loss of yield at the refinery in producing such fuels. CONCAWE published a study in 1983 (CONCAWE, 1983) stating that the optimum unleaded gasoline octane at the refinery would be 94.5 R+O. While higher octanes can provide better engine performance, however, with current refining technology, the demand of higher pool octane would require heavy capital investment at the refineries and decrease the gasoline yield per barrel of crude.

An alternative approach to this problem is to examine new technologies that could improve engine performance without the need of high octane fuels. Implementation of the pre-engine converter concept (Chen and Lucki, 1974) involving the concept of attaching a catalytic reactor to an internal combustion engine and converting a low octane liquid fuel to a high octane gas/liquid fuel, eliminates the problems associated with producing high performance fuels at the refinery, and shifts the task of reducing overall energy consumption back to the engine builders.

The concept of attaching a catalytic reactor to an internal combustion engine is not new. Cook (1940) disclosed a method of converting liquid hydrocarbons to gaseous fuel on board an engine. On-board production of hydrogen by steam reforming of gasoline fuel was studied by Newkirk and Abel (1972). However, the practicality of this concept using conventional catalytic systems in constrained in the past by the available space in a vehicle and the longevity of the catalyst. Medium pore zeolites, on the other hand, with their high activity and excellent stability have been demonstrated to fulfill these requirements (Chen and Lucki, 1974; Weisz et al., 1974; Chen, 1978).

Shown in Figure 8.4 are the data on the octane-boosting capability of ZSM-5 at 482°C as a function of space velocity.

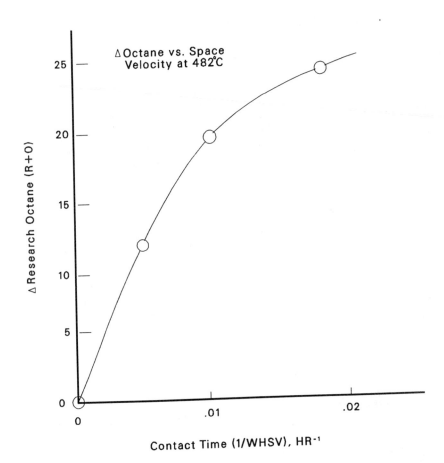

Figure 8.4   Octane vs. space velocity at 482°C.

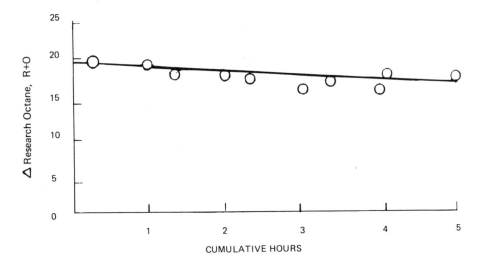

Figure 8.5   Octane vs. on-stream time at 482°C, 100 WHSV.

Figure 8.5 shows the stability of the catalyst. The catalyst retains its octane-boosting capability over 5 hours of on-and-off operation.

Implementation of such a system, however, still awaits the development of advanced engine technology to satisfy such requirements as cold start, catalyst regeneration, and drive-ability. Recent advances in engine technology, such as computer control of air-fuel ratio, spark timing (Ikeura et al., 1980), and knock sensors (Currie et al., 1979) have gone a long way toward making this concept practical.

VI. LIQUID FUEL FROM BIOMASS

A. Direct Conversion of Carbohydrates

A number of technologies for the conversion of biomass to liquid fuels are currently available. They include, among others, fermentation, pyrolysis, hydrogenative liquefaction and two-stage conversion processes such as gasification followed by alcohol synthesis or Fischer-Tropsch synthesis. Unfortunately, they either require large capital investments or are high in energy consumption and low in net carbon recoveries (Weisz, 1982; Weisz and Marshall, 1984; Chen et al., 1986).

A significant improvement in the energy efficiency of any two-stage conversion process is possible if the two separate stages can be combined into a single stage such that the endothermic gasification reaction is coupled with the exothermic CO reduction reaction. In this single-stage process, the removal of oxygen from the biomass occurs simultaneously with the enrichment of the hydrogen content of the liquid product. Furthermore, an improvement in the recovery of carbon is possible if the overall carbon-to-hydrogen ratio of the products can be reduced to below 2 (Chen et al., 1986).

An embryo of such a single-stage process is indicated by the catalytic conversion of sugars and hydrolyzed over HZSM-5. As was discussed in Chapter 4, when aqueous solutions of sugars and hydrolyzed starch were passed over HZSM-5 at $510°C$, both hydrocarbons and CO were produced directly from these materials combining the endothermic and exothermic reactions of a two-stage process. Examination of the data in Table 4.28 and Figure 4.24 shows that in addition to the hydrocarbon products, large quantities of CO were produced, convertible to hydrocarbons through Fischer-Tropsch synthesis or methanol synthesis. When the CO with the hydrocarbons is counted, nearly 60% of the carbon in xylose is recovered as premium products, exceeding the net yield of current technologies.

Experiments with mixtures of carbohydrates and methanol showed that a synergism exists between carbohydrate and methanol, increasing the yield of hydrocarbons from the sugars (Figure 8.6). The best yield was with glucose: 48.7 wt % carbon as hydrocarbons, 13.7 wt % as CO, and 33.4% as coke, giving an estimated overall carbon recovery of 72.5%, significantly higher than current technologies.

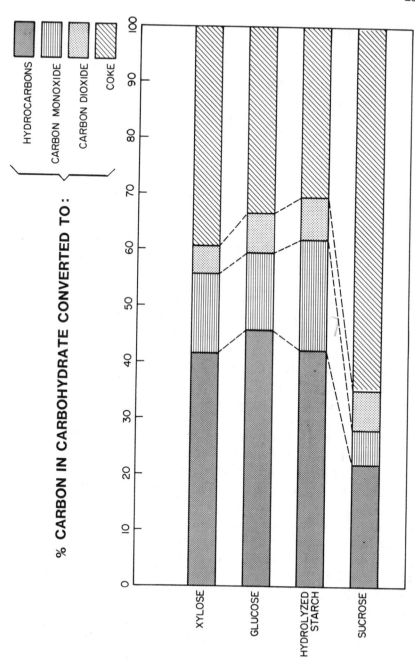

Figure 8.6 Production distribution from 4:1 methanol-carbohydrate mixtures at 510°C, 2 WHSV, 1 atm. (From Chen et al., 1986.)

Unlike the current technology which requires multistep processing, the direct conversion process is nearly isothermal and therefore requires little process energy. But, of course, this is only the beginning; the process still has a long way to go toward commercialization.

B. Biomass Pyrolysis

Pyrolysis of wood has been practiced for centuries to produce charcoal, wood alcohol, and other gas and liquid products. In more recent years, considerable international research efforts have been devoted to the development of methods to increase the yield of pyrolysis liquid (Antal, 1982; 1985) and to apply catalysis to upgrade the pyrolysis liquid to gasoline-like liquids (Chantal et al., 1984).

A number of new approaches to pyrolysis, including the use of short contact times (Scott et al., 1987; Diebold and Scahill, 1985; 1987a) and subatmospheric pressures (Lemieux et al., 1987), are being investigated. Primary pyrolysis product yields, as high as 75 wt % of the dry wood converted, have been reported (Diebold et al., 1986). These so-called "primary" pyrolysis products appear to be depolymerized fragments of the cellulose, hemicellulose, and lignin present in the original wood (Evans et al., 1984). Their ultimate analysis also resembles that of the original wood, which has an effective hydrogen index of near zero. However, they are highly reactive, forming tarry products upon standing if not further processed. The upgrading of these unstable products to stable and more valuable fuels and chemicals represents a new challenge to catalysis.

Attempts to convert pyrolysis liquids to gasoline-like hydro-carbons over ZSM-5 (Chantal et al., 1984, Dao et al., 1987; Diebold and Scahill, 1987b; Chen et al., 1987) have given disappointingly low yields to be of practical interest. This is to be expected by their low effective hydrogen index. Higher yields can be expected only if some of the oxygen is retained in the final product, or through selective decarbonylation and decarboxylation reactions to enrich the hydrogen content of the liquid product.

As discussed in Chapter 4, the conversion of other natural products, which have higher effective hydrogen indices than carbohydrates, such as corn oil, peanut oil, castor oil, and jojoba oil over ZSM-5, has been reported by Weisz et al. (1979).

VII. APPLICATIONS IN OTHER INDUSTRIES

A. Fermentation

A process similar to the MTG process converting ethanol to hydrocarbons may be integrated into the ethanol fermentation process to produce fuels and chemicals from biomass. This integration can bring significant savings in energy cost, because the energy-intensive rectifying columns may be eliminated by processing the overhead product from the beer column directly over the catalyst (Chen, 1983; Whitcraft et al., 1983; Aldridge et al., 1984; Anunziata et al., 1985).

Another possibility is to take advantage of the selective sorption properties of ZSM-5, which may be used to extract ethanol from the fermenter beer (Chen and Miale, 1983; Dessau, 1983; Dessau and Haag, 1984; Bul et al., 1985; Chen and Miale, 1985; Chen and Miale, 1987). By lowering the ethanol concentration in the fermenter with zeolites, it is conceivable that the rate of fermentation and the throughput of a fermenter can be significantly increased.

B. Chemicals

In addition to the already commercialized processes described in Chapter 6, a number of other commercially useful chemicals can be produced over the medium pore zeolites. By taking advantage of the selectivity and stability of the catalyst, improved yield and process economics over the conventional technology can be expected.

Among various hydrocarbon products, medium pore molecular sieves should find application in the production of durene (Chang et al., 1975b), cumene (Kaeding, 1983), and para-dialkylbenzenes such as p-diethylbenzene (Ishida and Nakajima, 1986) and p-di-isopropylbenzene (Kaeding, 1985). The selectivity and stability advantages may also be realized in the production of hetero-compounds, such as alcohols (Chang and Morgan, 1980), glycols (Chang and Hellring, 1986), methyl tert-butyl ether (Chu and Kuehl, 1987) and the synthesis of selected isomers of aromatic amines (Chang and Lang, 1983; 1984), pyridines (Chang and Lang, 1980), alkyl-pyridines (Chang and Perkins, 1983) and alkylphenols (Kaeding et al., 1980; Young and Burress, 1980; Wu, 1983).

The isomerization activity of the medium pore zeolites for a large variety of aromatic and heterocyclic compounds should find applications in the production of difficult-to-synthesize isomers. For example, the m-cresol could be easily produced from its o- and/or p-isomers by isomerizing the latter isomers over the medium pore zeolites.

A novel route for the manufacture of m-cresol and 1, 3-hydroxybenzene was proposed by Kaeding et al. (1980) starting with alkylation of toluene with propene to a mixture of meta/para-isomers, followed by the selective dealkylation of the para-isomer. The m-cymene is then converted to the final products by oxidation and rearrangement reactions, analogous to conditions for the commercial cumene-to-phenol process. The process avoids the cumbersome method of separating the m-isomer from an equilibrium cresol mixture.

Similar applications in the pharmaceutical industry can also be expected.

C. Polymer Waste Recovery

A new polymer waste recovery technology developed by Fuji Tech and Mobil has received some attention recently (CNN News, 1988). The process combines catalytic conversion over ZSM-5 with thermal depolymerization of polymer wastes. Its commercialization should open a new front for the application of shape selective catalysis.

REFERENCES

Aiura, M., T. Masunaga, K. Moriya, and Y. Kageyama, Fuel *63*, 1138 (1984).

Aldridge, G. A., X. E. Veryklos, and R. Mutharasan, Ind. Eng. Chem. Process Des. Dev. *23*, 733 (1984).

Anderson, J. R. and P. Tsai, Appl. Catal. *19*, 141 (1985).

Angevine, P. J., G. H. Kuehl, and S. Mizrahi, U.S. Pat. 4,431,518, Feb. 14, 1984.

CNN News, Mar. 8, 1988.

Anunziata, O. A., O. A. Orio, E. R. Herrero, A. F. Lopez, and A. R. Suarez, Appl. Catal. *15*, 235 (1985).

Antal, M. J., Jr., "Biomass Pyrolysis. A Review of the Literature. Part 1: Carbohydrate Pyrolysis," in *Advan. Solar Energy*, *1*, 61 (1982), K. W. Boer and J. A. Duffie, eds., Am. Solar Energy Soc., Boulder, Colo.

Antal, M. J., Jr., "Biomass Pyrolysis. A Review of the Literature. Part 2: Lignocellulose Pyrolysis," in *Advan. Solar Energy*, *2*, 175 (1985), K. W. Boer and J. A. Duffie, eds., Am. Solar Energy Soc./Plenum Press, New York.

Brophy, J. H., J. J. Fontfreide, J. D. Tomkinson, U.S. Pat. 4,652,688, Mar. 24, 1987.

Bul, S., X. Veryklos, and R. Mutharasan, Ind. Eng. Chem. Process Des. Dev. *24*, 1209 (1985).

Butter, S. A., A. T. Jurewicz, and W. W. Kaeding, U.S. Pat. 3,894,107, July 8, 1975.

Chang, C. D., W. H. Lang, and A. J. Silvestri, U.S. Pat. 3,894,104, Aug. 9, 1975a.

Chang, C. D., A. J. Silvestri, and R. L. Smith, U.S. Pat. 3,894,105, Aug. 9, 1975b.

Chang, C. D. and W. H. Lang, U.S. Pat. 4,220,783, Sept. 2, 1980.

Chang, C. D. and N. J. Morgan, U.S. Pat. 4,214,107, July 22, 1980.

Chang, C. D. and W. H. Lang, U.S. Pat. 4,380,669, Apr. 19, 1983.

Chang, C. D. and P. D. Perkins, U.S. Pat. 4,388,461, June 14, 1983; U.S. Pat. 4,395,554, July 26, 1983.

Chang, C. D. and W. H. Lang, U.S. Pat. 4,434,299, Feb. 28, 1984.

Chang, C. D. and S. D. Hellring, U.S. Pat. 4,620,044, Oct. 28, 1986.

Chantal, P., S. Kaliaguine, J. L. Grandmaison, and A. Mahey, Appl. Catal. 10, 317 (1984).

Chen, N.Y. and S. J. Lucki, "Approaches to Automotive Emissions Control," R. W. Hurn, ed., Am. Chem. Soc. Symp. Ser. 1, 69 (1974).

Chen, N.Y., U.S. Pat. 4,070,993, Jan. 31, 1978.

Chen, N.Y. and D. S. Shihabi, U.S. Pat. 4,269,697, May 26, 1981.

Chen, N.Y., Chemtech 13, 488 (1983).

Chen, N.Y. and J. N. Miale, U.S. Pat. 4,420,561, Dec. 13, 1983.

Chen, N.Y. and J. N. Miale, U.S. Pat. 4,515,892, May 7, 1985.

Chen, N.Y., T. F. Degnan, Jr., and L. R. Koenig, Chemtech 16, 506 (1986).

Chen, N.Y. and J. N. Miale, U.S. Pat. 4,690,903, Sept. 1, 1987.

Chen, N.Y., D. E. Walsh, and L. R. Koenig, "Fluidized Bed Upgrading of Wood Pyrolysis Liquids and Related Compounds," Paper 67, Symp. on Cellulose and Related Materials, Am. Chem. Soc. 193rd Nat. Mtg., Denver, Apr. 5-10, 1987.

Chu, P. C. and G. H. Kuehl, Ind. Eng. Chem. Res. 26, 365 (1987).

CONCAWE Report, "Assessment of the Energy Balances and Economic Consequences of the Reduction and Elimination of Lead in Gasoline," CONCAWE, The Hague, Dec., 1983.

Cook, J. T., U.S. Pat. 2,201,965, May 21, 1940.

Currie, J. H., D. S. Grossman, and J. J. Gumbleton, "Energy Conservation with Increased Compression Ratio and Electronic Knock Control," Paper 790173, SAE-Mtg., Detroit, Feb. 26–Mar. 2, 1979.

Dao, L. H., M. Haniff, A. Houle, and D. Lamothe, "Reactions of Biomass Pyrolysis Oils Over ZSM-5 Zeolite Catalysts," Paper 71, Symp. on Cellulose and Related Materials, Am. Chem. Soc. 193rd Nat. Mtg., Denver, Apr. 5-10, 1987.

Dessau, R. M., U.S. Pat. 4,423,280, Dec. 27, 1983.

Dessau, R. M. and W. O. Haag, U.S. Pat. 4,442,210, Apr. 8, 1984.

Diebold, J. P. and J. W. Scahill, "Entrained-Flow Fast Ablative Pyrolysis of Biomass," Annual Report, Oct. 1, 1983–Nov. 30, 1984, Solar Res. Inst., Golden, Colo., SERI/PR-2665, 1985.

Diebold, J. P., H. L. Chum, R. J. Evans, T. A. Milne, T. B. Reed, and J. W. Scahill, "Low-Pressure Upgrading of Primary Pyrolysis Oils from Biomass and Organic Wastes," paper presented at the 10th IGT/CBETS Conf. on Energy from Biomass and Wastes, Washington, April 7–10, 1986.

Diebold, J. P. and J. W. Scahill, "Production of Primary Pyrolysis Oils in a Vortex Reactor," Paper 10, Symp. on Cellulose and Related Materials, Am. Chem. Soc. 193rd Nat. Mtg., Denver, Apr. 5–10, 1987a.

Diebold, J. P. and J. W. Scahill, "Biomass to Gasoline (BTG): Upgrading Pyrolysis Vapors to Aromatic Gasoline with Zeolite Catalysts at Atmospheric Pressure," Paper 70, Symp. on Cellulose and Related Materials, Am. Chem. Soc. 193rd Nat. Mtg., Denver, Apr. 5–10, 1987b.

Evans, R. J., T. A. Milne, and M. N. Soltys, J. Anal. Appl. Pyrol. 6, 273 (1984).

Gorring, R. L. and R. L. Smith, U.S. Pat. 4,153,540, May 8, 1979.

Haxbun, E. A., U.S. Pat. 4,556,749, Mar. 12, 1985.

Ikeura, K., A. Hosaka, and T. Yano, "Microprocessor Control Brings About Better Fuel Economy with Good Driveability," SAE Paper 800056, Detroit, Feb. 25–29, 1980.

Ishida, H. and H. Nakajima, U.S. Pat. 4,613,717, Sept. 23, 1986.

Jones, C. A., J. J. Leonard, and J. A. Sofranko, U.S. Pat. 4,567,307, Jan. 28, 1986.

Kaeding, W. W., M. M. Wu, L. B. Young, and G. T. Burress, U.S. Pat. 4,197,413, Apr. 8, 1980.

Kaeding, W. W., Eur. Pat. 148,584, July 17, 1985; Eur. Pat. 149,508, July 24, 1985.

Kaeding, W. W., U.S. Pat. 4,393,262, July 12, 1983.

Kaiser, S. W., U.S. Pat. 4,677,243, June 30, 1987.

Lemieux, R., C. Roy, B. de Caumia, and D. Blanchette, "Preliminary Engineering Data for Scale-Up of a Biomass Vacuum Pyrolysis Reactor," Paper 9, Symp. on Cellulose and Related Materials, Am. Chem. Soc. 193rd Nat. Mtg., Denver, Apr. 5–10, 1987.

LaPierre, R. B., R. L. Gorring, and R. L. Smith, "Extended End-Point Distillate Fuels from Shale Oil by Hydrotreating Coupled with Catalytic Dewaxing," paper presented at the Am. Chem. Soc. Natl. Mtg., New York, Apr. 17, 1986.

McCaslin, J. C., ed., *International Petroleum Encyclopedia*, Vol. 19, PennWell, Tulsa, Okla., p. 294 (1986).

Miller, W., M. Harvey, and M. Hunter, "Premium Syncrude from Oil Shale Using Union Oil Technology," paper presented at the NPRA Annual Mtg., San Antonio, March 1982.

Mitarai, Y., "Catalyst Research for Solvent Hydrotreating in Nedol Process," Paper B-51, 3rd China-Japan-USA Symp. on Catal., Xiamen, Aug. 7-11, 1987.

Newkirk, M. S. and J. L. Abel, U.S. Pat, 3,682,142, Aug. 8, 1972.

Pace Synthetic Fuels Report *24*(1), 4-72, J. E. Sinor, ed., Pace Consultants, Houston (March 1987).

Scott, D. S., J. Piskorz, A. Grinshpun, and R. G. Graham, "Effect of Temperature on Liquid Product Composition from the Fast Pyrolysis of Cellulose," Paper 7, Symp. on Cellulose and Related Materials, Am. Chem. Soc. 193rd Nat. Mtg., Denver, Apr. 5-10, 1987.

Shihabi, D. S., U.S. Pat. 4,284,529, Aug. 18, 1981.

Shihabi, D. S., U.S. Pat. 4,377,469, Mar. 3, 1983.

Sofranko, J. A., J. J. Leonard, C. A. Jones, A. M. Gaffney, and H. P. Withers, "Catalytic Oxidative Coupling of Methane over Sodium-Promoted Managnese Oxide on Silica and Magnesia," Paper C-8, 10th N. Am. Mtg. Catal. Soc., San Diego, May 17-22, 1987.

Walsh, D. E., U.S. Pat. 4,428,826, Jan. 31, 1984.

Ward, J. W. and T. L. Carlson, U.S. Pat. 4,600,497, July 15, 1986.

Weisz, P. B., N.Y. Chen, and S. J. Lucki, U.S. Pat. 3,855,980, Dec. 24, 1974.

Weisz, P. B., W. O. Haag, and P. G. Rodewald, Science *206*, 57 (1979).

Weisz, P. B., Chemtech *12*, 114 (1982).

Weisz, P. B. and J. F. Marshall, *Fuels from Biomass: A Critical Analysis of Technology and Economics*, Marcel Dekker, New York, 1984.

Whitcraft, D. R., X. E. Veryklos, and R. Mutharasan, Ind. Eng. Chem. Process Des. Dev. *22*, 452 (1983).

Whitehurst, D. D., ACS Symp. Ser. *139*, 133 (1980).
Whitehurst, D. D., T. O. Mitchell, and M. Farcasiu, *Coal Liquefaction*, Academic Press, New York, p. 274, 1984.
Wise, J. J., N. Y. Chen, and J. R. Katzer, paper presented at Am. Chem. Soc. Mtg., New York, Apr. 1986.
Wu, M. M., U.S. Pat. 4,391,998, July 5, 1983.
Yao, T., K. Hayakawa, K. Jurachi, and T. Chikada, Japan Pat. 61-181890, Aug. 14, 1986a.
Yao, T., K. Hayakawa, and K. Kurachi, Japan Pat. 61-203197, Sept. 9, 1986b.
Young, L. B. and G. T. Burress, U.S. Pat. 4,205,189, May 27, 1980.

9
Conclusions

Shape selective catalysis has made giant strides since its inception almost 30 years ago. From the original discovery that selective conversion of linear molecules can be achieved by molecular-size exclusion of nonlinear molecules from the catalytic sites, more subtle ways of achieving catalytic selectivity have been devised making use of such phenomena as configurational diffusional constraints and spatiospecific or transition state inhibition. With the discovery of new catalytic materials, such as the medium-pore zeolites, catalysts have been designed that can discriminate not only linear molecules from branched molecules but also molecules of varying degrees of chain branching and certain isomers of aromatic and naphthenic molecules, many of which are of industrial importance.

Industrial catalytic processing employing shape-selective catalysts are now a reality in petroleum refining and petrochemical industries. New catalytic dewaxing processes have created such new products as extended-boiling-range premium jet fuels and diesel fuels, and ultralow temperature paraffinic refrigeration oils and lubricating oils. Catalytic lube dewaxing process is replacing the solvent dewaxing process for its cost effectiveness in energy saving and operating simplicity. Operating in conjunction with current reforming and catalytic cracking processes, new processes using shape selective catalysts have broken the octane barriers of these traditional processes for producing gasoline, and offer new alternative processing schemes to refine crudes that will make more efficient use of hydrogen and increase a refiner's ability to handle hydrogen-deficient heavy crudes.

The design of shape selective aromatics processing catalysts have created new processing routes to BTX from a variety of starting materials including light gases. Several alumina-based and aluminum-chloride-based catalysts are being replaced by shape selective catalysts for their superior activity, stability and product selectivity in the production of xylenes and ethyl benzene, major starting materials for the synthetic fiber and polymer industries.

Similar advances have been made in the application of shape selective catalysis to the production of synthetic fuels and chemicals. Notable among them are the methanol-to-gasoline (MTG) process, the methanol-to-olefin (MTO) process and the demonstrated ability to produce a large variety of hydrocarbon and non-hydrocarbon chemicals.

With the ever-expanding knowledge of the synthesis and characterization of novel new materials, opportunities for new catalytic chemistry and new industrial applications can continue to be expected.

Author Index

Subject Index